SOCIETAL IMPLICATIONS OF NANOSCIENCE AND NANOTECHNOLOGY

National Science Foundation

At the request of the National Science and Technology Council (NSTC), Subcommittee on Nanoscale Science, Engineering, and Technology (NSET), the National Science Foundation organized a workshop on September 28-29, 2000. This report incorporates the views of leading experts from academia, private sector and government expressed at the workshop.
NSTC is the principal means for the U.S. President to coordinate science, space and technology policies across the Federal Government.

National Science Foundation

SOCIETAL IMPLICATIONS OF NANOSCIENCE AND NANOTECHNOLOGY

edited by

Mihail C. Roco

and

William Sims Bainbridge

National Science Foundation, Arlington, VA, U.S.A.

Logistical, Editing and Management Assistance by:

*International Technology Research Institute(ITRI),
World Technology (WTEC) Division, Loyola College, Baltimore, MD, U.S.A.*

R.D. Shelton, *ITRI Director*
G.M. Holdridge, *WTEC Division Director and Series Editor*
R. Horning, *IT Director, Web Administrator*

KLUWER ACADEMIC PUBLISHERS
DORDRECHT / BOSTON / LONDON

A C.I.P. Catalogue record for this book is available from the Library of Congress.

ISBN 0-7923-7178-X

Published by Kluwer Academic Publishers,
P.O. Box 17, 3300 AA Dordrecht, The Netherlands.

Sold and distributed in North, Central and South America
by Kluwer Academic Publishers,
101 Philip Drive, Norwell, MA 02061, U.S.A.

In all other countries, sold and distributed
by Kluwer Academic Publishers,
P.O. Box 322, 3300 AH Dordrecht, The Netherlands.

Printed on acid-free paper

This report was prepared under the guidance of NSET. Any opinions, conclusions or
recommendations expressed in this material are those of the authors and do not necessarily reflect
the views of the United States Government.

Printed in the Netherlands

SOCIETAL IMPLICATIONS OF NANOSCIENCE AND NANOTECHNOLOGY

Table of Contents

6.2 Focus on economic and political implications of potential technology

6.3 Focus on science and education implications

6.4 Focus on medical, environmental, space exploration and national security implications

6.5 Focus on social, ethical, legal, international and national security implications

EXECUTIVE SUMMARY

Advances in nanoscience and nanotechnology promise to have major implications for health, wealth, and peace in the upcoming decades. Knowledge in this field is growing worldwide, leading to fundamental scientific advances. In turn, this will lead to dramatic changes in the ways that materials, devices, and systems are understood and created. The National Nanotechnology Initiative (NNI) seeks to accelerate that progress and to facilitate its incorporation into beneficial technologies. Among the expected breakthroughs are orders-of-magnitude increases in computer efficiency, human organ restoration using engineered tissue, "designer" materials created from directed assembly of atoms and molecules, and the emergence of entirely new phenomena in chemistry and physics.

The study of the societal implications of nanotechnology must be an integral part of the NNI. An interagency effort within the U.S. Government, the NNI supports a broad program of nanoscale research in materials science, physics, chemistry, and biology; it explicitly seeks to create new opportunities for interdisciplinary work. It is balanced across five broad activities: fundamental research; grand challenges; centers and networks of excellence; research infrastructure; and, the ethical, legal, and social implications, including educational and workforce programs.

This report outlines some potential areas for research into societal implications of nanotechnology. It has been prepared just as the NNI is commencing, when there is greater opportunity to affect the NNI investment strategy. Research on societal implications will boost the chances for NNI's success and help the nation take advantage of new technology sooner, better, and with greater confidence. Moreover, sober, technically competent research on the interactions between nanotechnology and society will help mute speculative hype and dispel some of the unfounded fears that sometimes accompany dramatic advances in scientific understanding.

Toward this end, the National Science and Technology Council (NSTC), Committee on Technology (CT), Subcommittee on Nanoscale Science, Engineering, and Technology (NSET) — the Federal interagency group coordinating the NNI — sponsored a workshop on "Societal Implications of Nanoscience and Nanotechnology." Held September 28–29, 2000 at the National Science Foundation, this workshop brought together nanotechnology researchers, social scientists, and policy makers representing academia, government, and the private sector. Their charge was to: (1) survey current studies on the societal implications of nanotechnology (educational, technological, economic, medical, environmental, ethical, legal, etc.); (2) identify investigative and assessment methods for future studies of societal implications; (3) propose a vision for accomplishing nanotechnology's promise while minimizing undesirable consequences.

This report sponsored by NSF incorporates fully the views, opinions and presentations contributed by workshop participants and other leading experts. The NSET report to the NSTC Committee on Technology presents a more concise perspective. The workshop participants offered recommendations to: (a) accelerate the beneficial use of nanotechnology while diminishing the risks, (b) improve research and education, and (c) guide the contributions of key organizations. These recommendations, summarized below, serve as a basis for both the NNI participants and the public to begin addressing societal issues of nanotechnology:

- Make support for social and economic research studies on nanotechnology a high priority. Include social science research on the societal implications in the nanotechnology research centers, and consider creation of a distributed research center for social and economic research. Build openness, disclosure, and public participation into the process of developing nanotechnology research and development program direction.

- The National Nanotechnology Coordination Office (NNCO) should establish a mechanism to inform, educate, and involve the public regarding potential impacts of nanotechnology. The NNCO should receive feedback from the nanotechnology community, social scientists, the private sector, and the public with the goals of (a) continuously monitoring the potential societal opportunities and challenges; and (b) providing timely input to responsible organizations.

- Create the knowledge base and institutional infrastructure to evaluate nanotechnology's scientific, technological, and societal impacts and implications from short-term (3 to 5 year), medium-term (5 to 20 year), and long-term (over 20 year) perspectives. This must include interdisciplinary research that incorporates a systems approach (research-technology development-societal impacts), life cycle analysis, and real time monitoring and assessment.

- Educate and train a new generation of scientists and workers skilled in nanoscience and nanotechnology at all levels. Develop specific curricula and programs designed to:

a. introduce nanoscale concepts into mathematics, science, engineering, and technological education;

b. include societal implications and ethical sensitivity in the training of nanotechnologists;

c. produce a sufficient number and variety of well-trained social and economic scientists prepared to work in the nanotechnology area;

d. develop effective means for giving nanotechnology students an interdisciplinary perspective while strengthening the disciplinary expertise they will need to make maximum professional contributions; and

e. establish fruitful partnerships between industry and educational institutions to provide nanotechnology students adequate experience with nanoscale fabrication, manipulation, and characterization techniques.

- Encourage professional societies to develop forums and continuing education activities to inform, educate, and involve professionals in nanoscience and nanotechnology.

Over the next 10 to 20 years, nanotechnology will fundamentally transform science, technology, and society. However, to take full advantage of opportunities, the entire scientific and technology community must set broad goals; creatively envision the possibilities for meeting societal needs; and involve all participants, including the general public, in exploiting them.

1. INTRODUCTION

A revolution is occurring in science and technology, based on the recently developed ability to measure, manipulate and organize matter on the nanoscale — 1 to 100 billionths of a meter. At the nanoscale, physics, chemistry, biology, materials science, and engineering converge toward the same principles and tools. As a result, progress in nanoscience will have very far-reaching impact.

The nanoscale is not just another step toward miniaturization, but a qualitatively new scale. The new behavior is dominated by quantum mechanics, material confinement in small structures, large interfacial volume fraction, and other unique properties, phenomena and processes. Many current theories of matter at the microscale have critical lengths of nanometer dimensions. These theories will be inadequate to describe the new phenomena at the nanoscale.

As knowledge in nanoscience increases worldwide, there will likely be fundamental scientific advances. In turn, this will lead to dramatic changes in the ways materials, devices, and systems are understood and created. Innovative nanoscale properties and functions will be achieved through the control of matter at its building blocks: atom-by-atom, molecule-by-molecule, and nanostructure-by-nanostructure. Nanotechnology will include the integration of these nanoscale structures into larger material components, systems, and architectures. However, within these larger scale systems the control and construction will remain at the nanoscale.

Today, nanotechnology is still in its infancy, because only rudimentary nanostructures can be created with some control. However, among the envisioned breakthroughs are orders-of-magnitude increases in computer efficiency, human organ restoration using engineered tissue, "designer" materials created from directed assembly of atoms and molecules, as well as emergence of entirely new phenomena in chemistry and physics.

Nanotechnology has captured the imaginations of scientists, engineers and economists not only because of the explosion of discoveries at the nanoscale, but also because of the potential societal implications. A White House letter (from the Office of Science and Technology Policy and Office of Management and Budget) sent in the fall of 2000 to all Federal agencies has placed nanotechnology at the top of the list of emerging fields of research and development in the United States. The National Nanotechnology Initiative was approved by Congress in November 2000, providing a total of $422 million spread over six departments and agencies.

Nanotechnology's relevance is underlined by the importance of controlling matter at the nanoscale for healthcare, the environment, sustainability, and almost every industry. There is little doubt that the broader implications of this nanoscience and nanotechnology revolution for society at large will be profound.

National Nanotechnology Initiative

The National Nanotechnology Initiative (NNI, http://nano.gov) is a multi-agency effort within the U.S. Government that supports a broad program of Federal nanoscale research in materials, physics, chemistry, and biology. It explicitly seeks to create opportunities for interdisciplinary work integrating these traditional disciplines. The NNI will accelerate the pace of fundamental research in nanoscale science and engineering, creating the knowledge needed to enable technological innovation, training the workforce needed to exploit that knowledge, and providing the manufacturing science base needed for future commercial production. Potential breakthroughs are possible in areas such as materials and manufacturing, medicine and healthcare, environment and energy, biotechnology and agriculture, electronics and information technology, and national security. The effect of nanotechnology on the health, wealth, and standard of living for people in this century could be at least as significant as the combined influences of microelectronics, medical imaging, computer-aided engineering, and man-made polymers developed in the past century.

The NNI is balanced across five broad activities: fundamental research; grand challenges; centers and networks of excellence; research infrastructure; and societal/workforce implications. Under this last activity, nanotechnology's effect on society – legal, ethical, social, economic, and workforce preparation – will be studied to help identify potential concerns and ways to address them. As the NNI is commencing, there is a *rare opportunity to integrate the societal studies and dialogues from the very beginning and to include societal studies as a core part of the NNI investment strategy.*

NSET Workshop on "Societal Implications of Nanoscience and Nanotechnology"

Research on societal implications will boost the NNI's success and help us to take advantage of the new technology sooner, better, and with greater confidence. Toward this end, the National Science and Technology Council (NSTC), Committee on Technology (CT), Subcommittee on Nanoscale

Science, Engineering, and Technology (NSET) — the Federal interagency group that coordinates the NNI — sponsored a workshop on "Societal Implications of Nanoscience and Nanotechnology." Held September 28–29, 2000 at the National Science Foundation, this workshop brought together nanotechnology researchers, social scientists, and policy makers representing academia, government, and the private sector. It had four principal objectives:

- Survey current studies on the societal implications of nanotechnology (educational, technological, economic, medical, environmental, ethical, legal, cultural, etc.).

- Identify investigative and assessment methods for future studies of societal implications.

- Propose a vision and alternative pathways toward that vision integrating short-term (3 to 5 year), medium-term (5 to 20 year), and long-term (more than 20 year) perspectives.

- Recommend areas for research investment and education improvement.

This report addresses issues far broader than science and engineering, such as how nanotechnology will change society and the measures to be taken to prepare for these transformations. The conclusions and recommendations in this report will provide a basis for the NNI participants and the public to address future societal implications issues.

Chapters 2 through 5 of this report present the conclusions and recommendations that arose from the workshop. The participants' statements on societal implications are in Chapter 6, and a list of participants and contributors is in Appendix A. Selected endorsements of the NNI are provided as a reference (Appendix B).

2. NANOTECHNOLOGY GOALS

Nanoscale science and engineering will lead to better understanding of nature; advances in fundamental research and education; and significant changes in industrial manufacturing, the economy, healthcare, and environmental management and sustainability. Examples of the promise of nanotechnology, with projected total worldwide market size of over $1 trillion annually in 10 to 15 years, include the following:

3

- *Manufacturing*: The nanometer scale is expected to become a highly efficient length scale for manufacturing once nanoscience provides the understanding and nanoengineering develops the tools. *Materials* with high performance, unique properties and functions will be produced that traditional chemistry could not create. Nanostructured materials and processes are estimated to increase their market impact to about $340 billion per year in the next 10 years (Hitachi Research Institute, personal communication, 2001).

- *Electronics*: Nanotechnology is projected to yield annual production of about $300 billion for the semiconductor industry and about the same amount more for global integrated circuits sales within 10 to 15 years (see R. Doering, page 84-93 of this report).

- *Improved Healthcare*: Nanotechnology will help prolong life, improve its quality, and extend human physical capabilities.

- *Pharmaceuticals*: About half of all production will be dependent on nanotechnology — affecting over $180 billion per year in 10 to 15 years (E. Cooper, Elan/Nanosystems, personal communication, 2000).

- *Chemical Plants*: Nanostructured catalysts have applications in the petroleum and chemical processing industries, with an estimated annual impact of $100 billion in 10 to 15 years (assuming a historical rate of increase of about 10% from $30 billion in 1999; "NNI: The Initiative and Its Implementation Plan," page 84).

- *Transportation*: Nanomaterials and nanoelectronics will yield lighter, faster, and safer vehicles and more durable, reliable, and cost-effective roads, bridges, runways, pipelines, and rail systems. Nanotechnology-enabled aerospace products alone are projected to have an annual market value of about $70 billion in ten years (Hitachi Research Institute, personal communication, 2001).

- *Sustainability*: Nanotechnology will improve agricultural yields for an increased population, provide more economical water filtration and desalination, and enable renewable energy sources such as highly efficient solar energy conversion; it will reduce the need for scarce material resources and diminish pollution for a cleaner environment. For example, in 10 to 15 years, projections indicate that nanotechnology-based lighting advances have the potential to reduce worldwide consumption of energy by more than 10%, reflecting a savings of $100 billion dollars per year and a corresponding reduction of 200 million tons of carbon emissions ("NNI: The Initiative and Its Implementation Plan," page 93).

Knowledge and Scientific Understanding of Nature

The study of nanoscale systems promises to lead to fundamentally new advances in science and engineering and in our understanding of biological, environmental, and planetary systems. It also will redirect our scientific approach toward more generic and interdisciplinary research. Nanoscience is at the unexplored frontiers of science and engineering, and it offers one of the most exciting opportunities for innovation in technology.

Nanotechnology will provide the capacity to create affordable products with dramatically improved performance. This will come through a basic understanding of ways to control and manipulate matter at the nanometer scale and through the incorporation of nanostructures and nanoprocesses into technological innovations. It will be a center of intense international competition when it lives up to its promise as a generator of technology.

Nanotechnology promises to be a dominant force in our society in the coming decades. Commercial inroads in the hard disk, coating, photographic, and pharmaceutical industries have already shown how new scientific breakthroughs at this scale can change production paradigms and revolutionize multibillion-dollar businesses. However, formidable challenges remain in fundamental understanding of systems on this scale before the potential of nanotechnology can be realized.

Today, nanotechnology is still in its infancy, and only rudimentary nanostructures can be created with some control. The science of atoms and simple molecules, on one end, and the science of matter from microstructures to larger scales, on the other, are generally established. The remaining size-related challenge is at the nanoscale — roughly between 1 and 100 molecular diameters — where the fundamental properties of materials are determined and can be engineered. A revolution has been occurring in science and technology, based on the recently developed ability to measure, manipulate and organize matter on this scale. Recently discovered organized structures of matter (such as carbon nanotubes, molecular motors, DNA-based assemblies, quantum dots, and molecular switches) and new phenomena (such as giant magnetoresistance, coulomb blockade, and those caused by size confinement) are scientific breakthroughs that merely hint at possible future developments.

The nanoscale is not just another step toward miniaturization, but a qualitatively new scale. The new behavior is dominated by quantum mechanics, material confinement in small structures, large interfaces, and other unique properties, phenomena and processes. Many current theories of matter at the microscale have critical lengths of nanometer dimensions; these theories will be inadequate to describe the new phenomena at the nanoscale.

Nanoscience will be an essential component in better understanding of nature in the next decades. Important issues include greater interdisciplinary research collaborations, specific education and training, and transition of ideas and people to industry.

Industrial Manufacturing, Materials and Products

The potential benefits of nanotechnology are pervasive, as illustrated in the fields outlined below.

Nanotechnology is fundamentally changing the way materials and devices will be produced in the future. The ability to synthesize nanoscale building blocks with precisely controlled size and composition and then to assemble them into larger structures with unique properties and functions will revolutionize materials and manufacturing. Researchers will be able to develop material structures not previously observed in nature, beyond what classical chemistry can offer. Some of the benefits that nanostructuring can bring include lighter, stronger, and programmable materials; reductions in life-cycle costs through lower failure rates; innovative devices based on new principles and architectures; and use of molecular/cluster manufacturing, which takes advantage of assembly at the nanoscale level for a given purpose.

The Semiconductor Industry Association (SIA) has developed a roadmap for continued improvements in miniaturization, speed, and power reduction in information processing devices — sensors for signal acquisition, logic devices for processing, storage devices for memory, displays for visualization, and transmission devices for communication. The SIA roadmap projects the future of nanoelectronics and computer technology to approximately 2010 and to 0.1 micron (100 nanometer) structures, just short of fully nanostructured devices. The roadmap ends short of true nanostructured devices because the principles, fabrication methods, and techniques for integrating devices into systems at the nanoscale are generally unknown. New approaches such as chemical and biomolecular computing, and quantum computing making use of nanoscale phenomena and nanostructures, are expected to emerge.

The molecular building blocks of life — proteins, nucleic acids, lipids, carbohydrates, and their non-biological mimics — are examples of materials that possess unique properties determined by their size, folding, and patterns

6

at the nanoscale. Biosynthesis and bioprocessing offer fundamentally new ways to manufacture chemicals and pharmaceutical products. Integration of biological building blocks into synthetic materials and devices will allow the combination of biological functions with other desirable materials properties. Imitation of biological systems provides a major area of research in several disciplines. For example, the active area of bio-mimetic chemistry is based on this approach.

Medicine and the Human Body

Living systems are governed by molecular behavior at the nanometer scale, where chemistry, physics, biology, and computer simulation all now converge. Recent insights into the uses of nanofabricated devices and systems suggest that today's laborious process of genome sequencing and detecting the genes' expression can be made dramatically more efficient through use of nanofabricated surfaces and devices. Expanding our ability to characterize an individual's genetic makeup will revolutionize diagnostics and therapeutics. Beyond facilitating optimal drug usage, nanotechnology can provide new formulations and routes for drug delivery, enormously broadening the drugs' therapeutic potential.

Increasing nanotechnological capabilities will also markedly benefit basic studies of cell biology and pathology. As a result of the development of new analytical tools capable of probing the world of the nanometer, it is becoming increasingly possible to characterize the chemical and mechanical properties of cells (including processes such as cell division and locomotion) and to measure properties of single molecules. These capabilities complement (and largely supplant) the ensemble average techniques presently used in the life sciences. Moreover, biocompatible, high-performance materials will result from the ability to control their nanostructure. Artificial inorganic and organic nanoscale materials can be introduced into cells to play roles in diagnostics (e.g., quantum dots in visualization), but also potentially as active components. Finally, nanotechnology-enabled increases in computational power will permit the characterization of macromolecular networks in realistic environments. Such simulations will be essential for developing biocompatible implants and for studying the drug discovery process. An open issue is how the healthcare system would change with such large changes in medical technology.

Sustainability: Agriculture, Water, Energy, Materials, and Clean Environment

Nanotechnology will lead to dramatic changes in the use of natural resources, energy, and water, as outlined in the following paragraphs. Waste and pollution will be minimized. Moreover, new technologies will allow recovery and reuse of materials, energy, and water.

Environment

Nanoscience and engineering could significantly affect molecular understanding of nanoscale processes that take place in the environment; the generation and remediation of environmental problems through control of emissions; the development of new "green" technologies that minimize the production of undesirable by-products; and the remediation of existing waste sites and streams. Nanotechnology also will afford the removal of the smallest contaminants from water supplies (less than 200 nanometers) and air (under 20 nanometers) and the continuous measurement and mitigation of pollution in large areas.

In order to hasten the integrated understanding of the environmental role of nanoscale phenomena, scientists and engineers studying the fundamental properties of nanostructures will need to work together with those attempting to understand complex processes in the environment. Model nanostructures can be studied, but in all cases the research must be justified by its connection to naturally occurring systems or to environmentally beneficial uses. Environments for investigations are not limited and might include terrestrial locations such as acid mines, subsurface aquifers, or polar environments.

Energy

Nanotechnology has the potential to significantly impact energy efficiency, storage, and production. Several new technologies that utilize the power of nanostructuring, but developed without benefit of the new nanoscale analytical capabilities, illustrate this potential:

- Increasing the efficiency of converting solar energy into useful forms.

- High efficiency fuel cells, including hydrogen storage in nanotubes.

- A long-term research program in the chemical industry on the use of crystalline materials as catalyst supports has yielded catalysts with well-defined pore sizes in the range of 1 nanometer to reduce energy consumption and waste; their use is now the basis of an industry that exceeds $30 billion a year ("NNI: The Initiative and Its Implementation Plan," page 84).

- Developed by the oil industry, the ordered mesoporous material MCM-41 (known also as "self-assembled monolayers on mesoporous supports," SAMMS), with pore sizes in the range of 10–100 nanometers, is now widely used for the removal of ultrafine contaminants (see work performed at Pacific Northwest National Laboratory in *Nanotechnology Research Directions,* Kluwer Academic Publishers, 2000, pp. 216-218).

- Several chemical manufacturing companies are developing a nanoparticle-reinforced polymeric material that can replace structural metallic components in automobiles; widespread use of those nanocomposites could lead to a reduction of 1.5 billion liters of gasoline consumption over the life of one year's production of vehicles, thereby reducing carbon dioxide emissions annually by more than 5 billion kilograms ("NNI: The Initiative and Its Implementation Plan," page 88).

- Significant changes in lighting technologies are expected in the next ten years. Semiconductors used in the preparation of light emitting diodes (LEDs) for lighting can increasingly be sculpted on nanoscale dimensions. In the United States, roughly 20% of all electricity is consumed for lighting, including both incandescent and fluorescent lights. In 10 to 15 years, projections indicate that such nanotechnology-based lighting advances have the potential to reduce worldwide consumption of energy by more than 10%, reflecting a savings of $100 billion dollars per year and a corresponding reduction of 200 million tons of carbon emissions ("NNI: The Initiative and Its Implementation Plan," pages 92 - 93).

- The replacement of carbon black in tires by nanometer-scale particles of inorganic clays and polymers is a new technology that is leading to the production of environmentally friendly, wear-resistant tires.

Water

Global population is increasing while fresh water supplies are decreasing. The United Nations predicts that by the year 2025 that 48 countries will be short of fresh water accounting for 32% of the world's population ("NNI: The Initiative and Its Implementation Plan", page 95). Water purification and desalinization are some of the focus areas of preventative defense and environmental security since they can meet future water demands globally.

Consumptive water use has been increasing twice as fast as the population and the resulting shortages have been worsened by contamination. Nanotechnology-based devices for water desalinization have been designed to desalt sea water using at least 10 times less energy than state-of–the art reverse osmosis and at least 100 times less energy than distillation. The critical experiments underpinning these estimations are underway now. This energy-efficient process is possible by fabricating of very high surface area electrodes that are electrically conductive using aligned carbon nanotubes, and by other innovations in the system design.

Agriculture

Nanotechnology will contribute directly to advancements in agriculture in a number of ways: (1) molecularly engineered biodegradable chemicals for nourishing plants and protecting against insects; (2) genetic improvement for animals and plants; (3) delivery of genes and drugs to animals; and (4) nano-array-based technologies for DNA testing, which, for example, will allow a scientist to know which genes are expressed in a plant when it is exposed to salt or drought stress. The application of nanotechnology in agriculture has only begun to be appreciated.

Space Exploration

The stringent fuel constraints for lifting payloads into earth orbit and beyond, and the desire to send spacecraft away from the sun for extended missions (where solar power would be greatly diminished) compel continued reduction in size, weight, and power consumption of payloads. Nanostructured materials and devices promise solutions to these challenges. Nanostructuring is also critical to the design and manufacture of lightweight, high-strength, thermally stable materials for aircraft, rockets, space stations, and planetary/solar exploratory platforms. The augmented utilization of miniaturized, highly automated systems will also lead to dramatic improvements in manufacturing technology. Moreover, the low-gravity, high-vacuum space environment may aid the development of nanostructures and nanoscale systems that cannot be created on Earth.

National Security

Defense applications include (1) continued information dominance through advanced nanoelectronics, identified as an important capability for the military; (2) more sophisticated virtual reality systems based on nanostructured electronics that enable more affordable, effective training; (3) increased use of enhanced automation and robotics to offset reductions in military manpower, reduce risks to troops, and improve vehicle performance; (4) achievement of the higher performance (lighter weight, higher strength) needed in military platforms while providing diminished failure rates and lower life-cycle costs; (5) needed improvements in chemical/biological/nuclear sensing and in casualty care; (6) design improvements in systems used for nuclear non-proliferation monitoring and management; and (7) combined nanomechanical and micromechanical devices for control of nuclear defense systems.

In many cases economic and military opportunities are considered to be complementary. Strong applications of nanotechnology in other areas would provide support for national security in the long term, and vice versa.

Moving into the Market

Since economists have not yet really begun research on nanotechnology, their insights are somewhat tentative and based on experience with earlier technologies. A common paradigm is that new applications will be initially more costly than existing technologies, but will achieve better performance. However, completely new technologies may be cheaper, such as chemical manufacturing to mass produce nanoelectronic circuits as opposed to current methods using lithography in microelectronics. Overall, nanotechnology will offer substantial advantages, being smaller, faster, stronger, safer, and more reliable. At the same time, it will require investments in new production facilities and in a host of ancillary industries supplying the raw materials, components, and manufacturing machines. Because it will take time to achieve economies of scale and to develop the most efficient fabrication methods, costs are likely to be relatively high in the beginning.

For this reason, nanotechnology-based goods and services will probably be introduced earlier in those markets where performance characteristics are especially important and price is a secondary consideration. Examples are medical applications and space exploration. The experience gained will reduce technical and production uncertainties and prepare these technologies for deployment into the market place. Similarly, in the private sector, technology transfer is likely to occur from performance-oriented areas (such as medicine) to price-oriented ones (such as agriculture). As a given technology matures, its cost may decline, leading to greater penetration of the market even where performance is not decisive.

The displacement of an old technology by a new one tends to be both slow and incomplete. Displacement of older methods will accelerate to the extent that nanotechnology extends its technical range and perhaps lowers its relative price. However, nanotechnology also is likely to stimulate innovations in older technologies that make them better able to compete — an ironic but potentially beneficial second-order effect.

The diffusion and impact of nanotechnology will be partly a function of the development of complementary technologies and of a network of users. Whole new industries may have to be developed — along with the trained scientists and technicians to staff them. There may be many obstacles along this road that ordinary market processes cannot easily overcome. An important role for the government will be to invest in the long-term, high-risk, high-gain research needed to create these new industries and to ensure that they are consistent with broader societal objectives.

Federal support of the nanotechnology initiative is necessary to enable the United States to take advantage of this strategic technology and remain competitive in the global marketplace well into the future. Focused research programs on nanotechnology have been initiated in other industrialized countries. Currently, the United States has a lead on synthesis, chemicals, and biological aspects; it lags in research on nanodevices, production of nano-instruments, ultra-precision engineering, ceramics, and other structural materials. Japan has an advantage in nanodevices and consolidated nanostructures; Europe is strong in dispersions, coatings, and new instrumentation.

3. NANOTECHNOLOGY AND SOCIETAL INTERACTIONS

The Interactive Process of Innovation and Diffusion

New technologies come into being through a complex interplay of technical and social factors. The process of innovation that will produce nanotechnology and diffuse its benefits into society is complex and only partially understood. Economists, as well as scholars in other fields, have long studied the generation, diffusion, and impact of scientific and technological innovation. These studies outline the variables likely to determine the rate and direction of these impacts, and to identify relevant research questions. They provide a foundation on which to build studies of societal implications of nanotechnology.

Scientific discoveries do not generally change society directly; they can set the stage for the change that comes about through the confluence of old and new technologies in a context of evolving economic and social needs. The thorough diffusion of even major new developments rarely happens all at once. Nanotechnologies are so diverse that their manifold effects will likely take decades to work their way through the socio-economic system. While market factors will determine ultimately the rate at which advances in nanotechnology get commercialized, sustained support for nanoscience research is necessary in this early stage of development so as not to become a rate-limiting factor. Expediting research (innovation) and its incorporation into beneficial technology is a major challenge to the NNI.

Unintended and Second-order Consequences

Perhaps the greatest difficulty in predicting the societal impacts of new technologies has to do with the fact that once the technical and commercial feasibility of an innovation is demonstrated, subsequent developments may be as much in the hands of users as in those of the innovators. The diffusion and impact of technological innovations often depends on the development of complementary technologies and of the user network. As a result, new technologies can affect society in ways that were not intended by those who initiated them. Often these unintended consequences are beneficial, such as spin-offs with valuable applications in fields remote from the original innovation. For instance, consider how the Internet has progressed from a technology supported by the Department of Defense's Advanced Research Projects Agency (DARPA) to facilitate digital communications among universities with DARPA contracts, to a means by which teenagers and college students exchange music files. In another example, intended benefits may also have unintended or "second-order consequences." Nanotechnology-based medical treatments, for example, may significantly improve life span and quality of life for elderly people; a second-order consequence would be an increase in the proportion of the population that is elderly, which might require changes in pensions or health insurance, an increase in the retirement age, or a substantial increase in the secondary careers undertaken by older people. Another potential consequence that would need to be addressed is the potential increase of inequality in the distribution of wealth that we may call the "nano divide." Those who participate in the "nano revolution" stand to become very wealthy. Those who do not may find increasingly difficult to afford the technological wonders that it engenders. One near-term example will be in medical care: nanotech-based treatments may be initially expensive, hence accessible only to the very rich. Other consequences are not so desirable, such as the risk of closing old industries and environmental pollution, which sometimes becomes a problem, especially for large-scale technologies.

To assess a nanotechnology (or any technology) in terms of its unintended consequences, researchers must examine the entire system of which the technology is a part through its entire life cycle. As the case of electric automobiles illustrates, without a careful analysis of the entire set of activities that produce, operate, and eventually dispose of a technology, people may leap to false conclusions about the extent to which the technology pollutes. For example, manufacture and disposal of an electric vehicle's battery may release more lead into the environment than if the vehicle had been fueled throughout its working life by leaded gasoline.

One concern about nanotechnology's unintended consequences raises the question of the uncontrolled development of self-replicating nanoscale machines. A number of very serious technical challenges would have to be overcome before it would be possible to create nanoscale machines that could reproduce themselves in the natural environment. Some of these challenges appear to be insurmountable with respect to chemistry and physical principles, and it may be technically impossible to create self-reproducing mechanical nanoscale robots of the sort that some visionaries have imagined. A new form of life different from that known (i.e., carbon-based) would be a dramatic change that is not foreseen in the near future.

Initially, the impact of nanotechnology will likely be limited to a few specific products and services. Nanotechnology-based goods and services will probably be introduced earlier to those markets where consumers are willing to pay a premium for new or improved performance. Such primary effects would be to make things work better, cheaper, with more features, etc. This might, for example, increase food yields, generate new textiles for clothing, improve power production, or cure a certain disease. As mentioned above, by and large, the displacement of an old technology by a new one tends to be both slow and incomplete. As a result, nanotechnology will coexist for a long time with older technologies rather than suddenly displacing them. During that time it will affect the further development of those competing technologies. Other secondary effects might be shifts in demand for products and services, so that people come to expect different kinds of food, medical care, entertainment, etc. This shift in demand may also initiate a tertiary effect, the need for augmented nanotechnology infrastructure — interdisciplinary research centers, new educational programs to supply nanoscientists and nanotechnologists, etc. Other tertiary effects would move upstream in our social structures and cultural patterns, such as shifts in education and career patterns, family life, government structure, and so forth. *While there is no way of knowing, a priori, the unintended and higher order consequences of nanotechnology, the participation of social scientists in the NNI may allow for important issues to be identified earlier, the right questions to be raised, and necessary corrective actions taken.*

An effective and cost-efficient way to protect the public and deal with nanotechnology's potential negative consequences is to develop a tradition of social-science-based countermeasures — and to support research in publicly recognized institutions on the processes that develop nanotechnology and apply it in diverse areas of life.

Ethical Issues and Public Involvement in Decision Making

An important aim of a societal impact investigation of nanotechnology is to identify harms, conflicts over justice and fairness, and issues concerning respect for persons. For example, changes in workforce needs and human resources are likely to bring benefits to some and harm to others. Other examples of potential issues include safeguards for workers engaged with hazardous production processes, equity disputes raised by intellectual property protection, and questions about relationships between government, industry, and universities.

Scientists and engineers bring to their work a laudable concern for the social value of their labors. However, those working in a particular technical field may be focused on the immediate technical challenges and not see all of the potential social and ethical implications. It is important to include a wide range of interests, values, and perspectives in the overall decision process that charts the future development of nanotechnology. Involvement of members of the public or their representatives has the added benefit of respecting their interests and enlisting their support.

The inclusion of social scientists and humanistic scholars, such as philosophers of ethics, in the social process of setting visions for nanotechnology is an important step for the NNI. As scientists or dedicated scholars in their own right, they can respect the professional integrity of nanoscientists and nanotechnologists, while contributing a fresh perspective. Given appropriate support, they could inform themselves deeply enough about a particular nanotechnology to have a well-grounded evaluation. At the same time, they are professionally trained representatives of the public interest and capable of functioning as communicators between nanotechnologists and the public or government officials. Their input may help maximize the societal benefits of the technology while reducing the possibility of debilitating public controversies.

15

In addition, attention needs to be given to the individual responsibility of engineers, scientists, and others involved in the processes of generating powerful new nanotechnologies. Professional societies have a role to play in providing opportunities for discussing and devising guidelines that incorporate relevant ethical principles into emerging issues. Perhaps most importantly, ethics must be incorporated effectively into the curriculum for training new nanoscientists, nanotechnologists, and nanofabrication technicians.

Education of Nanoscientists, Nanotechnologists, and Nanofabrication Technicians

The United States faces the daunting challenge of attracting enough of the best graduate students to the physical sciences and engineering disciplines. Under present conditions, far too few good students are attracted to the fields relevant to nanotechnology. To some extent, this is a problem faced by all of the sciences, but the problem is especially acute for nanotechnology because a very large number of talented scientists, engineers, and technicians will be needed to build the nanotechnology industries of the future, and these professionals will require an interdisciplinary perspective.

Development of nanotechnology will depend upon multidisciplinary teams of highly trained people with backgrounds in biology, medicine, applied and computational mathematics, physics, chemistry, and in electrical, chemical, and mechanical engineering. Team leaders and innovators will probably need expertise in multiple subsets of these disciplines, and all members of the team will need a general appreciation of the other members' fields. Developing a broadly trained and educated workforce presents a severe challenge to our four-year degree and two-year degree educational institutions, which favor compartmentalized learning. Because current educational trends favor specialization, there must be fundamental changes in our educational systems. However, introducing new degree programs in nanotechnology that provide a shallow overview of many disciplines, none in sufficient depth to make major contributions, may not give students the training that is needed to meet the future challenges. The right balance between specialization and interdisciplinary training needs to be worked out through innovative demonstration programs and research on the education process and workforce needs.

Education in nanoscience and nanotechnology requires special laboratory facilities that can be quite expensive. Given the cost of creating and sustaining such facilities, their incorporation into nanotechnology workforce development presents a considerable challenge. Under the present education system, many engineering schools, let alone the two-year-degree colleges, cannot offer students any exposure to the practice of nanofabrication. Innovative solutions will have to be found, such as new partnerships with industry and the establishment of nanofabrication facilities that are shared by consortia of colleges, universities, and engineering schools. Web-based, remote access to those facilities may provide a powerful new approach not available previously.

Despite the tremendous educational challenges, the exciting intellectual, economic, and social opportunities of nanotechnology might become a major factor in reinvigorating our nation's youth for careers in science and technology.

Education of Social Scientists

A related educational challenge is the very small number of social scientists who have the technical background and research orientation that would allow them to conduct competent research on the societal implications of nanotechnology. At the university level, liberal arts education gives far too low a priority to scientific literacy. Social science professional societies, universities, and government agencies will have to make a long-term commitment to attract talented young social scientists to this area of research and to encourage them to gain the necessary professional skills and awareness of nanotechnology. This will require research on the societal implications of nanotechnology at a consistent and high enough level to establish this as a viable field of social science research.

4. SOCIAL SCIENCE APPROACHES FOR ASSESSING NANOTECHNOLOGY'S IMPLICATIONS

Social Science Research Approaches and Methodologies

It is important to have social scientists study the processes by which nanoscience is conducted and nanotechnology is developed — even at this early stage. The knowledge gained will help policymakers and the public understand how nanoscience and nanotechnology are advancing, how those advances are being diffused, and how to make necessary course corrections. Insight into the innovation process will also grow.

Social scientists and scholars possess many effective ways of studying the development of new technology and its implications for society. Some

17

methodologies suitable for studying nanotechnology are known; others will have to be identified or developed. Ethnographic techniques, such as those traditionally employed by anthropologists, are appropriate for some of this work. Also useful will be interviews of research and development teams, conducted over time and augmented by surveys and historical methods, to document the evolution of the knowledge and technology. Interviews, social network techniques, studies of communication patterns, and citation analysis of publications more generally can offer insights into the diffusion of scientific discoveries and ideas. Application of a scientific idea to a technical problem, technology transfer, and introduction of products into the marketplace can be tracked through statistics on research and development investments, patent applications, and new products and services.

With concerted effort, it will be possible to develop a number of indicators that provide early signs of change. One challenge to social science research will be to identify "bellwethers," "early adopters," or "first movers." For example, some geographic areas and strata of society experience technological change earlier than others do. Incipient transformations may reveal themselves first in start-up companies, university labs, and Internet communications.

Innovating activity takes place in academe, industry, government laboratories, Federal agencies, and professional societies. For each group, measures and methodologies for studying the process and content of change must be developed. In academe, key indicators might include interdisciplinary work, new courses, fellowships, information flow, and regional coalitions. In the private sector, key indicators might include investments, startups, and corporate partnerships. For government laboratories, data would cover budgets, equipment, standards, and coalitions, and for agencies, examination might be given to new initiatives, databases, and centers. The professions will create some number of new forums, symposiums, journals, and job fairs where interdisciplinary topics and careers would flourish. Social science research areas relevant to the process of discovery, invention, and development include appearance of new ideas and innovations, change in societal goals, and shift in commercial investment.

The societal impacts of nanotechnology may be of great scope and variety. A second research challenge is to address both short- and long-term impacts, intended and unintended, and first-order through Nth-order consequences. Because nanotechnology presents a highly diverse set of novel technical possibilities, accurate prediction of even the immediate consequences of individual innovations may be impossible. Some impacts will be surprising, and others will have emergent implications that will reveal themselves only over a long period of time. Ultimately, both technology and society are elaborate systems with the potential for chaotic and variable feedback mechanisms. Nanotechnology has such promise to impact so many aspects of society that predictions will be uncertain and difficult to validate empirically. This observation should not discourage researchers, however, but should inspire them to invest considerable sophistication and effort in their work. The domains and measures of potential social impacts include: economic growth, employment statistics, social transformations and medical statistics.

A third challenging, but important, area for social science research is the social acceptance, resistance, or rejection of nanotechnology. Representative sample surveys, supplemented by focus groups and open-ended interviews, can measure affective, cognitive, and psychosocial parameters. In recent years, political scientists and sociologists have developed new computation-intensive techniques for studying coverage by the news media; they have been tested in research on public controversies and are ready to track the changing public perceptions of nanotechnology. These and more traditional methods can also be applied to charting the process of regulatory review and approval, court decisions that actively sanction the use of the technology, mobilization of political support and opposition, and the activities of relevant social movements. There are multiple feedback loops in which society responds to new innovations and in so doing transforms the context in which innovation occurs. As more and more new nanotechnologies are publicized and actually appear in the marketplace, the variable degree of social acceptance will become ever more important. Indicators to measure social acceptance of nanotechnology will be needed in the following areas: economic, political, religious and cultural.

Institutional Infrastructure for Societal Implications Research

Nanotechnology's vast scope and the necessity of bringing together researchers from different disciplines may require that some of the important social science research be carried out by large teams, housed in research centers set up for this purpose. At the same time, many of the research methodologies require social scientists to be on site where nanoscientists, nanotechnologists, and decision-makers are doing their work. One model that meets both of these requirements is the virtual distributed research center (VDRC). Under this approach, each VDRC would be organized around a specific but somewhat broad set of scientific questions and research methods, so that the members would have a common framework for designing, carrying out, and communicating their research. To ensure that results reflect the wide diversity of nanotechnology, social scientists would have to examine a range of empirical settings — for example, by conducting ethnographic research in a variety of nanotechnology laboratories. Thus many individual members of the VDRC would be situated, or would spend large blocks of time, at the geographically dispersed sites where they are studying. However, the VDRC would have a physical center that coordinates the work, develops and maintains funding and institutional partnerships, and supports effective communications among the far-flung team, both electronically and face-to-face in periodic meetings.

Many important nanotechnology-related questions could best be examined by more traditional centers, teams, and individual investigators. For example, survey research on public attitudes might best be done by a conventional team of researchers connected to one of the existing social survey organizations. Some research on economic trends, changing labor markets, and publication patterns could be done by individual investigators with access to data already available. Finally, there will always be a need for innovative projects carried out by individual scientists or small teams to develop new theories and methodologies and to carry out reconnaissance studies of emerging social phenomena.

5. RECOMMENDATIONS

Given the tremendous potential benefits of nanotechnology, and the concern that it be developed with sensitivity to potential negative implications, the workshop participants offered the following recommendations:

- Make support for social and economic research studies on nanotechnology a high priority. Include social science research on the societal implications in the nanotechnology research centers, and consider creation of a distributed research center for social and economic research. Build openness, disclosure, and public participation into the process of developing nanotechnology research and development program direction.

- The National Nanotechnology Coordination Office should establish a mechanism to inform, educate, and involve the public regarding potential impacts of nanotechnology. The mechanism should receive feedback from the nanotechnology community, social scientists, the private sector, and the public with the goals of (a) continuously monitoring the potential societal opportunities and challenges; and (b) providing timely input to responsible organizations.

- Create the knowledge base and institutional infrastructure to evaluate nanotechnology's scientific, technological, and societal impacts and implications from short-term (3 to 5 year), medium-term (5 to 20 year), and long-term (over 20 year) perspectives. This must include interdisciplinary research that incorporates a systems approach (research-technology development-societal impacts), life cycle analysis, and real time monitoring and assessment.

- Educate and train a new generation of scientists and workers skilled in nanoscience and nanotechnology at all levels. Develop specific curricula and programs designed to:

  (a) introduce nanoscale concepts into mathematics, science, engineering, and technological education;

  (b) include societal implications and ethical sensitivity in the training of nanotechnologists;

  (c) produce a sufficient number and variety of well-trained social and economic scientists prepared to work in the nanotechnology area;

  (d) develop effective means for giving nanotechnology students an interdisciplinary perspective while strengthening the disciplinary expertise they will need to make maximum professional contributions; and

  (e) establish fruitful partnerships between industry and educational institutions to provide nanotechnology students adequate experience with nanoscale fabrication, manipulation, and characterization techniques.

- Encourage professional societies to develop forums and continuing education activities to inform, educate, and involve professionals in nanoscience and nanotechnology.

Other Measures

- Involve social scientists at the onset of major nanotechnology R&D activities, while the technology is still in an early stage of development, from vision setting to development projects. Extend the NNI grand coalition of academe, the private sector, and government to include the social, behavioral and economic science communities. Coordinate this activity through the National Nanotechnology Coordination Office.

- Prepare corresponding management plans and policies to ensure that we can respond flexibly to implications as they appear on the horizon.

- Integrate short-, medium-, and long-term objectives and ensure intermediate outcomes.

Specific Areas for Research and Education Investment

- Invest in significant new innovative efforts to educate and train the nanoscience and nanotechnology workforce, including the sensitivity to societal implications and the introduction of nanoscale concepts in mathematics, science, engineering, and technological education. Conduct a comprehensive study to determine the distinctive educational and workforce issues related to nanotechnology and seek potential solutions for problems that are identified in that study.

- Support interdisciplinary research that includes a systems approach (research-technology-society), life cycle analysis, and real-time monitoring and assessment. Study the evolution of disruptive technologies, the winners and losers in major technological transformations, and the implications for the economy. From those studies, project the social purpose, social equity implications, and social enterprise dynamics of anticipated nanotechnologies

Recommendations to Organizations

Academe:

- Focus on multidisciplinary work on key research and education issues concerning socio-economic implications. Support interdisciplinary interactions between the physics, chemistry, biology, materials, and engineering communities on one hand, and the social sciences and economic science research communities on the other hand.

- Educate and train a new generation of scientists and workers for nanoscience and nanotechnology at all levels.

- Create local information centers for the public, teachers, industry, and scholars.

Private Sector:

- Provide intellectual input and seed funding of activities aimed at assessing the societal implications of nanotechnology.

- Develop partnerships with academic institutions and other sectors.

- Offer accessibility to social science researchers and provide feedback on societal implications studies.

Government R&D Laboratories:

- Establish interdisciplinary teams for major grand challenges in nanotechnology including socio-economic perspectives, including social scientists.

- Develop databases for evaluation and continuously update scenarios for the future.

- Establish user facilities available to industry and academe that enable integration of basic and applied research.

- Support nanotechnology research within laboratories emphasizing national defense mission.

Government Funding Agencies:

- Support nanotechnology researchers and social scientists to study the societal implications of nanotechnology.

- Support NNCO or an advisory group to monitor developments and examine the socio-legal implications of nanoscience and nanotechnology and take appropriate actions. Communicate the resulting activities to the public.

- Provide coordinated support for long-term basic research and shorter-term technological developments to create the technological base and prove the potential of the new technology.

- Establish dialog between NNCO and information technology (NSO) and biotechnology (BECON) coordinating groups elsewhere in the Federal Government.

Professional Societies:

- Develop forums and continuing education activities to inform, educate, and involve professionals and the public.

- Provide suggestions for grand challenges and suggest warning signs of potential risks.

With an Eye to the Future

Nanotechnology will fundamentally transform science, technology, and society. In 10 to 20 years, a significant proportion of industrial production, healthcare practice, and environmental management will be changed by the new technology. Economic growth, personal opportunities, sustainable development and environmental preservation will be affected. To take full advantage of the new technology, the entire scientific and technology community must involve all participants, including the general public; creatively envision the future; set broad goals; and work together to expedite societal benefits.

6. STATEMENTS ON SOCIETAL IMPLICATIONS

6.1 OVERVIEWS

NATIONAL NANOTECHNOLOGY INITIATIVE

T. Kalil, White House

(Transcript from September 28, 2000)

Good morning, it's a pleasure for me to be here. I want to recognize the work that Mike Roco and the other members of the Nanoscience & Technology Working Group have been putting into developing this initiative. What I am going to do this morning is just give you an overview of the initiative, and then talk a little bit about some of the issues that I hope you will address today.

The national nanotechnology initiative was unveiled by President Clinton in a speech that he gave on science and technology policy in January of 2000. He called for an initiative with funding levels around 500 million dollars, and as he noted some of the research goals may take 10, 20, or even more years for us to realize. I think that's very important because I think there is a tendency in a new field when there is a lot of excitement to over-promise. For example, artificial intelligence that led to the famous AI winter, when people said that a strong artificial intelligence was just around the corner. I think it's important that people — even though there's a huge amount of enthusiasm about this area — continue to give the public a sense of how long it will take to make some of these breakthroughs. This gives you a sense for what was in the President's budget.

We estimate that in FY 2000 government agencies were investing around 270 million dollars in nanotechnology, and the President proposes to roughly double that. As you know, congress has not concluded the appropriations process right now. There are a number of areas where we are very concerned that congress has not provided full funding for this. For example, in the NSF budget, which is really the lead agency for the initiative, the congress has provided 125 million dollars for nanotechnology compared to our request of 217 million dollars. That's below the level that NSF was proposing to put into nanotechnology even if they didn't get any budget increase. NSF thought this area was important enough so that they were going to reallocate some of their base funding. We are particularly concerned about that, but we are also concerned about the lack of full funding for the initiative in DOE and NIST as well.

The initiative had five elements. The first was to increase support for fundamental research. The second was to pursue a set of grand challenges, which I am going to talk about later. The third was to support a series of centers of excellence, primarily university-based. The fourth was to increase support for research infrastructure. The fifth, what we are engaged in today, is to think about the ethical, economic, legal and social implications and also address the education and training of the nanotechnology workforce.

So what are some examples of some of these grand challenges? One was the ability to have really dramatic improvements in our ability to store and process information the way the President talked about this in the CalTech speech was the ability the equivalent of the Library of Congress in a device the size of a sugar cube; developing nanoengineered MRI contrast agents that might allow us to detect tumors that were a few cells in size as opposed to waiting until they are visible to the human eye; materials that were ten times stronger than steel at a fraction of the weight; doubling the energy efficiency of solar cells from 20% to 40% and something that was of great interest to NASA — the ability to have a continuous unmanned presence outside the solar system. So those were just a few examples of the types of grand challenges that we thought were possible.

Why now did the administration decide that this was an area that really deserved a great deal of attention and additional resources? First of all, I think there was a sense within the science and engineering community that this was an area that could have huge potential payoff and could be a technology that is every bit as significant as electricity or the development of the transistor or the Internet. It was also an area where, clearly, long-term, high risk research is needed which is where there is an important government role. For research that has a payoff that is longer than five years, it's very difficult for individual companies to justify to their shareholders making those sorts of investments.

This was an area that was also clearly interdisciplinary, where it's going to require collaboration between the biological/physical sciences and engineering, and we also saw it as a way of increasing support for the physical sciences and engineering. The reason that that's important is that, although biomedical research has enjoyed strong support on the Hill, support for the physical sciences and engineering has been stagnant. That's a real problem, both because these are important disciplines in their own right, and also because if we hope to make progress in biomedical research it will draw on innovations that are coming from physical sciences and engineering. A communications problem that we've had is that while there were members of Congress like speaker Gingrich that understood intuitively why it was a good idea to spend more money on biomedical research. We did not enjoy that depth and breadth of support for the physical sciences and engineering. So, increasing funding in this area was one way of addressing that major challenge in science policy.

There was a high level of enthusiasm in the community. This is not something that was a top-down initiative. I think it was really driven by the fact that the funding agencies were getting many more meritorious proposals than they were able to fund. This is a time when we are limited by dollars rather than ideas. NSF, for example, even after limiting proposals to two per campus was only able to fund 12% of the proposals that were coming in. We had some early promising results. The HP, UCLA and other breakthroughs in molecular electronics is a good example of that. This was an area that was important for multiple agency missions. We need nanotechnology if we are going to stay on the Moore's Law curve of improvements in price and performance in computers and electronics. Finally, particularly if we want to prepare the workforce that is going to be needed to capitalize on these new technologies, we need to increase support particularly in our universities. So that is, sort of in a nutshell, why the administration decided to make this area a priority and an area of emphasis in the President's 2001 budget.

Now, in thinking about the societal implications of nanotechnology, I think that we have to acknowledge from the outset that this is a very difficult exercise. We are at a stage where this is an inherently speculative exercise. As some of the speakers noted in their presentations, nanotechnology is sort of an umbrella term for a wide range of technologies. We've got differences of opinion about what the ultimate outcome of nanotechnology research is going to be, which is to be expected given that it is still a very young and undefined field. Even when we have technologies that are widely diffused, such as information technology, you don't have any consensus on what the impact is. So, does that mean that we should just sort of wash our hands and say let's not even bother thinking about it? I don't think that's the right answer either.

I think there a couple of directions that we can move down that may be fruitful — one is to identify particular applications of nanotechnology that are going to be broader societal objectives in areas like environment and health. Second, I think we can try to determine what lessons can be learned from studying the impacts of other technologies. For example, if we look at the literature on the impact of IT, a recurring theme of information

27

technology and technology in general is something that is part of a broader socio-technical system. The reason that you can't predict what the impact is going to be is that it depends on the broader social, economic, and cultural context in which those technologies are embedded. I think that we need to start thinking of the potential risks and downsides. I think that although there are a number of points in Bill Joy's article that one could take issue with, I think it's difficult to deny his thesis that a lot of these technologies are going to end up increasing the destructive capability of small groups. Some of these technologies are not going to require Manhattan Project level of efforts in order to produce significant destructive capabilities. So, I think it behooves people to start thinking of those issues now as opposed to later. Then, I think you can engage in some scenario planning to say what are some different plausible scenarios. For example, what if we had 30 or 40 more years of Moore's Law style progress in storage and processing and what would that mean for our economy and our society? Even though this is an inherently difficult and speculative exercise, I don't think it's too early to start thinking about it.

I will conclude with a couple of thoughts. One is that I think we really need to reject this naïve technological determinism that I thought was best summed up by the slogan of the 1933 Chicago World's Fair, which is "science finds, industry applies, man conforms." I think that you encounter an attitude that technology is something that is totally out of control, and if it can happen it will happen. I think that's a dangerous attitude that we ought to reject. I think the other area that is particularly difficult in the area of nanotechnology is "keep an open mind but not so open that your brain falls out." I forget who said that, but I think one of the things that makes this discussion particularly difficult when you get into the thinking about 20 or 30 years out, is that different people have a different dividing line between keeping an open mind and allowing the brain to fall out. I think that's one of the things that will make this complex issue interesting. The third question I have is: there are certainly some people that believe that not only are we going to see continued change, which has been something that we are all familiar with, but that the rate of change itself is going to accelerate dramatically, so how seriously should we take this notion of some people in the field, particularly science fiction writers, that things are going to get very different and very weird over the next 20 to 40 years? So, with that I want to thank everyone for coming and participating in this workshop, because we really are going to need the best minds thinking and working in this issue. Thank you very much.

THE AGE OF TRANSITIONS

N. Gingrich, American Enterprise Institute

My perspective on the societal implications of nanoscience is as an historian, an amateur student of science and as an elected official with a long time in government. I have done an extensive amount of reading and talked to people all across America who have made or are on the edge of breakthroughs in science and technology. The newest, least understood, and most promising area of science, in my opinion is nano scale science and technology.

I'd like to try and put nanoscience in some kind of historical perspective, starting with the concept of an S-curve of technology (see Figure 6.1). (A more detailed version is the "Age of Transitions" at http://www.newt.org.)

Mature Flattening Out

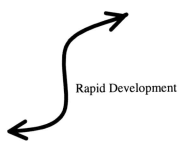

Rapid Development

Slow Build-up

Figure 6.1. The concept of an S-curve of technology.

As a general rule breakthroughs start relatively slow; they build up momentum, they suddenly reach a period of catalytic change, they go up the curve of capability very, very rapidly and then as they mature they tend to level out.

The S-curve we have been experiencing, the revolution in computing and communications that has been dubbed the Information Age, began around 1965. It is the result of two developments — computing and communications.

Computing is a key element in this change and we are only one-fifth of the way into developing the computer revolution. To take one metric, according to Professor James Meindl, the chairman of the Georgia Tech Microelectronics Department, the first computer built with a transistor was "Tradic" in 1955, and it had only 800 transistors. The Pentium II chip has 7.5 million transistors. The Pentium III chip has 29 million transistors. In the next year or so an experimental chip will be built with one billion transistors. Within fifteen to twenty years there will be a chip with one trillion transistors. Graphing that scale of change, it is enormous and its implications are huge.

Yet focusing only on computer power understates the rate of change. Communications capabilities are going to continue to expand dramatically and that may have as big an impact as computing power on our society and economy. Today most homes get Internet access at 28,000 to 56,000 bits per second. Within a few years a combination of new technologies for compressing information (allowing you to get more done in a given capacity) with bigger capacity (fiber optic and cable) and entirely new approaches (such as satellite direct broadcast for the Internet) may move household access up to at least six million bits per second, and some believe we may reach the 110 million bits needed for uncompressed motion pictures. An amazing range of opportunities have and will continue to open up as our communications capabilities continue to improve and grow.

When you look at the distance we have traveled in relation to computing and communications capabilities and the distance scientists are predicting we will travel in the next 20 years, I believe that we are only one-fifth of the way along that S-curve (see Figure 6.2).

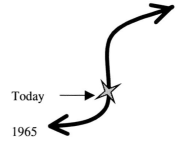

Computer and
Communications Revolution

Today

1965

Figure 6.2. Present position on the S-curve of technology.

30

Even that understates the rate and scale of change because there is a second S-curve that is beginning to develop, overlapping with the current S-curve. The best description I have found of this second wave of change is the NASA AMES laboratory version. In their mission statement, they use a triangle with biology on one side, nanoscience or nanotechnology on the other side, and information, by which they mean supercomputing and above, at the bottom of the triangle. It is the interaction of those three that, I think, leads to an enormous wave of change, which creates the second S-curve (see Figure 6.3 and Figure 6.4).

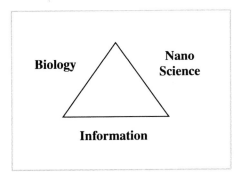

Figure 6.3.

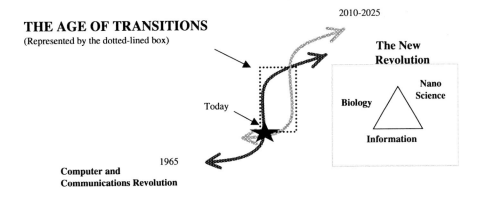

Figure 6.4. The age of transitions.

So, looking out over the next 30 years, as these two S-curves continue to accelerate and continue to overlap, we are going to be impacted by two large, profound waves of change. It is the overlapping period we are just beginning to enter that I believe will be an Age of Transitions.

The age of transitions will be an ever-evolving set of discoveries, with the information world increasingly interacting with the physical world.

The topic of this conference is the societal implications of nanoscience. Discoveries involving nanoscience will be as dramatic and, I believe, even more important than the creation of the Internet. Let's consider the economic impact nanoscience may have our society. Bill Joy, co-founder and Chief Scientist of Sun Microsystems, has estimated that the combination of the information and physical world will create in this century a thousand trillion dollars worth of wealth. As a former lawmaker, I thought I was used to dealing in big sums. This is really big! In fact, it would be adding 100 U.S. economies to the world market.

The application of nanoscience into nanotechnology will introduce disruptive technologies into our lives and, therefore, into the economy. Large corporations have been very successful at improving a product or service they are already providing. While working to improve existing products, their new science will also create disruptive technologies. It is very hard for corporations to incorporate disruptive technologies. Clayton M. Christensen, in *The Innovator's Dilemma: When New Technologies Cause Great Firms to Fail,* describes case after case of new entrants dominating a field by creating technology that big firms ignored. These disruptive technologies were initially too slow or too poor in performance to react well to meet the needs of the customers. It was not that these big corporations were stupid or irrational; their customers initially looked at the new technology and said, "I don't want it." A new layer of customers then appeared to create a new market that embraced the disruptive technology and it eventually became the norm. Disk drives, hydraulic shovels, mini-mills are just a few examples of technologies once considered disruptive and studied by Christensen.

When more research in the area of nanoscience is done, you will not only have a disruption in scientific assumptions that lead to new discoveries, you will see the nanoscience being applied to new economic sectors in ways that cannot be anticipated.

The application of nanoscience into real marketable products will be much more rapid than we saw with the Internet or other revolutionary technologies. When computing began, we had a remarkably primitive venture capital entrepreneurial system. As nanoscience is translated into nanotechnology it will be entering a very aggressive, very well financed, very experienced entrepreneurial venture world, which will be hungrily looking for the next big deal. So the rate of translation into new startups and new markets should be orders of magnitude faster than it was for computing.

Let's turn for a moment away from placing nanoscience in the context of historical trends and look more at the uniqueness of the science itself. As a student of science, I am going to make some assertions. These may be corrected by my other conference attendees but, again, as a student, the implications of nanoscience strike me as profound. Nanoscience is the base of how the world operates. There may be many layers below the atom, but for practical purposes an atom is an excellent base element. Taking the information nanoscience is teaching us about atoms and their activity and applying it has dramatic implications. Take the environment for example. We can learn to grow products using less energy and with less waste by product. It may mean that if we use dramatically less energy, then the projections and assumptions of current environmental debates, like Kyoto, are totally obsolete. It certainly means in biology that as we get better nano instrumentation and research tools, and if our nanoscale observational capabilities continue to grow, our capacity to deal with the complexities of human biology are going to go up dramatically and may not even be only orders of magnitude, but a different world of capabilities. For example, I recently met with the NSF cognitive science group and learned that in brain wave scanning, we are still at the molecular level. The potential of nano-level brain wave scanning is a new frontier in mind science. The application towards improving our education techniques alone are enormous.

If we want to stay at the forefront economically and remain a world leader politically and militarily, I think we have an obligation to really look seriously at funding more nanoscience research. Although President Clinton deserves credit for creating the National Nanotechnology Initiative, the amount he requested barely scratches the surface of what we need to be spending. This is an extraordinary national security issue. America is certainly not the only country working on advancing the field of nanoscience. The Japanese are working on it. The Europeans are working on it. In a decade you are going to see the Indians and the Chinese with a very major effort in this area because it will become obvious that it is so profoundly important.

After a year and a half of talking with scientists around the country I reached the conclusion that we need to rethink from the ground up how we design our science budget. In fact, I want to introduce the idea of an opportunities-based science budget (longer explanation can be found at www.newt.org.) If without any comprehensive effort you have eight times as many applications as you can finance (which is what many agency directors have testified), talking about a six percent or nine percent increase is inadequate.

I believe that we are actually at the edge of an age of discovery that is vastly richer than anybody yet understands. When I was at MIT they were very excited by the fact that the human ear has a million moving parts. We couldn't have discovered that fifteen or twenty years ago. We still don't fully understand what it means. Dr. Francis Crick, co-discoverer of the structure of the DNA molecule, told me that, in his judgment, it would take a hundred years of work to finish out the human genome project's implications. Most Americans think it's almost done, and the newspaper says they are almost done with "X" so we think we should relax. I am beginning to believe just the opposite. We are at the opening of an age of discovery. We have a whole new wave of things about to happen. Therefore, what I want to propose is an opportunities-based science budget. We actually go out and say to all scientists: if you had the money, if you were not in an economically constrained environment — what is it that we could learn across the whole system from astronomy to physics to math to biology to chemistry. I think what you would find is that we would have lots of grand projects we can't even dream of today because we start with a relatively limited budget. Again, I would cite the international geophysical year as a model that changed geology decisively because it was a large enough model that we had the data on the right level.

I have sent a letter to the appropriation chairmen and Senator Lott and Speaker Hastert suggesting that the real number for NSF (letter found at www.newt.org) ought to be to catch up with NIH. Since 1994, we've increased the NIH budget 72% and the NSF budget 27% — that is in the long run profoundly wrong because if you don't invest in math, physics, chemistry, etc., if you don't invest in basic research, basic instrumentation, you will run out of the capability to do fresh biological work. When Harold Varmus was the Director of NIH, he even testified to this point in front of Congress. It is vital that we reassert the centrality of fundamental research in this country.

Finally, we cannot lead the world if we do not profoundly overhaul math and science education. This is a sleeping crisis of unbelievable proportions. If, in fact, the scale of change is as large as I just suggested, then in order to stay at the forefront we have to have a lot more 19-year-olds capable of doing math and science. We just don't have them, and the current system isn't producing them. This is a crisis that requires at least an Eisenhower-level kind of national security look at this — even if that means paying high-schoolers to take calculus. It is literally worth our thinking through any change we have to make in order to produce a nation in which enough people are capable of doing math and science. Nothing less will do.

TECHNOLOGICAL IMPLICATIONS OF NANOTECHNOLOGY: WHY THE FUTURE NEEDS US

J.A. Armstrong, IBM VP, Sci. & Tech. (ret.)

Introductory Remarks

I have been asked to talk about "the technological implications of nanotechnology." This is a tall order for at least two reasons. First, because the technological and societal implications of the major, present nanotechnologies, semiconductor and magnetic recording, are so vast that it would take a month of workshops to explore them. Second, the topic is a tall order regarding the *new* nanotechnologies because no one is sure what they will be and which will be most successful and therefore pervasive and capable of having significant impacts, both good and (possibly) adverse.

I find that the very term "nanotechnology" — although wonderfully suited to the description of a welcome and significant funding initiative — is at too high a level of abstraction for our purposes here today. Which "nanotechnology" are we supposed to be talking about? Surely not semiconductors and magnetic recording, except as historical examples and sources of valuable lessons.

Do we mean the new nanotechnologies that make use of the fabrication methods of traditional silicon technology but extend them by incorporating exotic new materials from the realms of biology, chemical sensing, and genetic engineering?

Or do we mean technologies based on chemical and materials-science methods that can produce tiny particles — such as nanotubes and wires — with remarkable properties despite the lack of lithographically defined spatial ordering?

Or, as is likely, some hybrid of the above? Or something altogether different that we will hear about during the course of the workshop?

The question "Which nanotechnology?" is important because the societal impacts (almost certainly overwhelmingly benign, but possibly occasionally adverse) depend very much on which technology is involved, and *even more soon which application is involved.*

In their wisdom, the progenitors of this workshop have left all these questions open, no doubt with the intention of ensuring a very stimulating and wide ranging discussion. So I am going to put forth a list of five questions of my own, and then proceed to answer, or at least to address them.

Questions

First question: *Why are we having this workshop at all?* When the Administration and Congress fund an NSF initiative to build a high energy physics detector, or a supercomputer, or an Engineering Research Center, we do not normally proceed to collective scrutiny of possible societal impacts. May it be that we have promised too much in the way of a nano revolution, and aroused unease in the community at large? Is there some message that goes *beyond* nanoscience and technology that we should be alert to? Are more and more areas of scientific research going to be funded with these precautionary measures attached?

Second question: *What can we learn from the examples of past technology developments and societal impacts* that will be helpful in thinking about the future of nano developments? There is, in my view, much to be gained from reflecting on the emergence of semiconductor technology as a major force shaping society. Not all of the coming nanotechnologies will share attributes of the semiconductor revolution, but some will.

Third question: *Can we say anything useful* a priori *about the impacts of the **manufacturing** of new nanotechnological devices as distinct from the societal impacts of the **applications** to which the new devices will be put?* This is a subset of the previous question, but easier to deal with in concrete terms.

Fourth question: *How is one to measure societal impacts anyway?* (a) What will count as benign and what as adverse? Recall Joseph Schumpeter's characterization of the genius of modern capitalism as "creative destruction." Much of that creative destruction has been enabled by the digital revolution that in turn was made possible by nanotechnology. In many cases, one man's benign impact is another man's adverse impact. (b) Can we use one or more of the emerging nanotechnologies as test-beds for determining so-called "societal returns?" For example, what is to be counted in the set of societal returns, as opposed to the private returns which will accrue to firms that bring the new technologies to market? It is certain that the list of what is to be counted as "private return" is very different from the list of what is to be counted as societal return. Indeed, some of what is counted as societal return is counted by the private sector as *investment*, not return. (c) And how are we to measure adverse impacts quantitatively? I am neither an economist nor social scientist, but I am interested in these matters and frustrated by what I perceive as a lack of clarity and transparency in discussions by specialists.

Fifth question: In view of the miserable track record in long range forecasting that has been run up by scientific and technical experts over the years, *why would anyone take seriously what we have to say about societal impacts decades into the future* of any of the emerging new nanotechnologies?

Responses

The remainder of my talk will deal as fully with these questions as can be done in twenty minutes! But the main points I will try to make are these:

- We can say many plausible things about the possible evolution of new nanotechnologies. But because of the great uncertainties that surround the future, no **particular** view or concern about future impacts can have **scientific** claim to be so certain that policy should seriously constrain scientific options now.

- The whole aim of our forethought and intellectual preparation and policymaking should be to ensure that we can flexibly respond to impacts as they appear on the horizon, no matter how different they may be from what we expected.

- And therefore, the Future Does Need Us, we who can be flexible and rational and respond to surprises and unintended consequences, as well as respond to wonderful new opportunities. (If you are worried, as some seem to be, about a robotic future full of nano mechanisms that don't need us, I suggest you rent a copy of Woody Allen's *Sleeper* from the video store, and restore your sense of balance!)

DON'T COUNT SOCIETY OUT: A RESPONSE TO BILL JOY

J.S. Brown, Xerox Palo Alto Research Center; P. Duguid, University of California, Berkeley

Summary

The April issue of *Wired* carried an article by Bill Joy, cofounder and chief scientist of Sun Microsystems, called "Why The Future Doesn't Need Us."

The article argued that "our most powerful 21st-century technologies — robotics, genetic engineering, and nanotechnology — are threatening to make humans an endangered species." Here, we offer a response.

All of us need to worry about the concerns Joy raises. Technology is moving frighteningly fast. But much of the fear in Joy's article comes from a tendency among the digerati, when surveying technological change, to extrapolate from the steepest part of the curve or, in effect, to count in the order of 1, 2, 3, ... a million — or even infinity. You can see this in old predictions that a few years would take us from industrial nuclear power plants to domestic ones. And you can see it again in the short steps Joy takes from the possibility of replicating peptides to the imminent certainty of a robot society, or from the theory of nanotechnology to its practical implications.

This sort of counting is an example of what we call "tunnel vision." It excludes all the other factors that come into play as technologies develop. In particular, it excludes the social factors that always shape and redirect technology, making counting much harder. In making this argument, we are not arguing that there is therefore nothing to worry about. Far from it. The cause for worry is real. Instead, we are suggesting that — contrary to those who can only see disaster — something can be done. But what that something may be is very hard to see if tunnel vision cuts out all the forces in play except for the technological ones.

Our response asks that the social factors at work be factored in. Society and technology develop, we argue, in co-evolutionary steps, each profoundly affecting the other. When the social forces are left out of the picture, there seems nothing else to do but resign ourselves to wait for a future that doesn't need us. On the other hand, if the role of social forces in the co-evolutionary spiral is clear, then warnings like Joy's highlight the need to develop new social forms, new kinds of organization, and new formal and informal institutions to replace the slow, outmoded ones and to respond rapidly to rapid social change. It's at this level, we believe, that debate should be engaged.

Introduction

Whatever happened to the household nuclear power pack? The full-scale nuclear generator had barely left the drawing board before futurists predicted that every house would soon have a smaller version. From here, technoenthusiasts could see the end of power monopolies, the emergence of the "electronic cottage," the death of the city, and the long decline of the corporation. Pessimists and Luddites, of course, primarily foresaw localized nuclear meltdown and household nuclear weapons. Each side waited for Nirvana or Armageddon to roll by so it could triumphantly tell the other, "I told you so." They're still waiting.

Bill Joy's recent article "Why the Future Doesn't Need Us" (*Wired* 8.04) brings those old controversies to mind. In saying this, we do not want to underestimate the importance of Joy's much-cited article. No Luddite, Joy is an awe-inspiring technologist. So when he describes a technological juggernaut thundering towards society and worries that even those straight in its pathway are blindly cheering, all of us need to listen. Like the nuclear prognosticators, Joy can see the juggernaut clearly. Like them, too, he can't see any controls. Indeed, it's the absence of controls that makes his vision so scary. But it doesn't follow that the juggernaut is uncontrollable.

To understand why no controls are visible, readers should note the publication in which this article appears. For the best part of a decade, *Wired* has been an enjoyable cheerleader for the digital age. Its shift with Joy's article from cheering to warning marks an important moment in the digital *zeitgeist*. Finally, many prognosticators, like investors, are coming to realize that rapid technological innovation can have a down side. And as with many investors, the tone in *Wired* has swung straight from wild euphoria to high anxiety — as if there were no middle ground. That the change in mood should be so extreme is not all that surprising. When they felt we were all being triumphantly carried along by technology, the digerati saw little need to look for the brake. So now they fear that, rather than being carried, we are instead standing smack in technology's path, and they don't seem to know where a brake might be found.

To see where one might lie, let's go back to the nuclear power pack. Innovation, the argument went, would make nuclear plants smaller and cheaper. These would soon shrink to household size. Then they would enter mass production and quickly become available to all. The argument still seems unavoidable — until you notice what's missing. The tight focus of this vision makes it almost impossible to see forces other than technology at work. Yet in the case of nuclear development, there were many other forces at work. These included the environmental movement, anti-nuclear protests, concerned scientists, worried neighbors of Chernobyl and Three Mile Island, NIMBY responses to nuclear waste, government regulators, anti-proliferation treaties, and corporate-shareholder rebellions. Cumulatively, these forces slowed the nuclear juggernaut to a manageable crawl. Similar social forces are at work on modern technologies today. But because the digerati, like technoenthusiasts before them, look to the future through a narrow technological tunnel, they too have trouble bringing other forces into view.

The Tunnel Ahead

As an emblem of technological futurism, take the cover of Bill Gates's first book, *The Road Ahead*. This showed a smiling Gates standing before an empty blacktop stretching unproblematically into the future. There was the road. Gates pointed. We needed only to follow it. When the book appeared, *The Nation* magazine put an ad with this picture next to one for a Bruce Springsteen concert. In that, if memory serves, a world-weary Springsteen stood outside a tavern in some unidentified, industrial-age town. The tavern looked seedy. Traffic blocked the oily, rain-swept road. The guys outside the tavern probably weren't sober. And the women further down the road probably weren't waiting for buses. The contrast between the two ads reminded us how much easier it is to lay out the road ahead with confidence and plausibility if you only think about road and ignore all the messiness that people willfully bring to the picture.

Leaving people out of the picture and focusing on technology in splendid isolation, tunnel vision doesn't only lead to both exuberant and doom-and-gloom scenarios by the bucketful. It also leads to tunnel design — the design of "simple" technologies that are actually very difficult to use. So to escape both trite scenarios and bad design, we have to widen horizons and bring into view not just technological systems, but also social systems. Good designs look beyond the dazzling potential of the technology to social factors such as the limited patience of most users. Paying attention to the latter has, for example, allowed the Palm Pilot and Nintendo Gameboy to sweep aside more complex rivals. Their elegant simplicity has made them readily usable. And their usability has in turn created an important social support system. They are so widely used that now anyone having trouble using a Pilot or Gameboy rarely has to look far for a more experienced user to give advice.

As this small example suggests, technological and social systems shape each other. The same is true on a larger scale. Technologies, as gunpowder, the printing press, the railroad, the telegraph, and the Internet have shown, shape society in quite profound ways. But equally, social systems, in the form of government, the courts, formal and informal organizations, social movements, professional networks, local communities, market institutions, and so forth, shape, moderate, and redirect the raw power of technologies. The whole process might best be thought of as one of "co-evolution," with society and technology mutually shaping each other. In considering one, then, it's important to keep the other in mind. Given the crisp edges of technology and the fuzzy ones of society, it certainly isn't easy to grasp the two simultaneously. But grasp both you must, if you want to see where we are all going or design the means to get there.

Tidings of Discomfort?

This joint perspective allows a more sanguine look at the central concerns Bill Joy laid out in *Wired*: genetic engineering, nanotechnology, and robotics. Undoubtedly each deserves serious thought. But each should be viewed in the context of the social system in which it is inevitably embedded. That context provides, to return to our earlier metaphor, a glimpse of the brake and steering mechanisms on what otherwise would appear as an out-of-control juggernaut.

Genetic engineering presents the clearest example. Barely a year ago this seemed an unstoppable force. Major chemical and agricultural interests were barreling unstoppably along an open highway. In the past twelve months or so, however, road conditions have changed dramatically. Cargill has faced Third World protests against its patents. Monsanto has suspended research on sterile seeds. And champions of genetically modified foods, who once saw an unproblematic and lucrative future, are scurrying to counter consumer boycotts of their products. If, as some people fear, genetic engineering represents one of the horses of the Apocalypse, it is certainly no longer unbridled.

Erratic biotech stocks suggest that it's now very hard to see beyond this immediate, sharp curve in what once looked like an open road. There's no clear consensus — only a lot of name calling (Frankenfood! Luddites!). Almost certainly, those who support genetic modification will have to look beyond the labs and the technology if they want to advance. They need to address society directly — not just by labeling modified foods, but by educating people about the costs and the benefits. Of course, having ignored social concerns, proponents have made the people they need to educate profoundly suspicious and hostile. In consequence, they have made their road significantly more uphill.

Nanotechnology offers a rather different example of how the future can frighten us. For this, which involves engineering at a molecular level, both the promise and the threat seem unmeasurable. But they are unmeasurable for a good reason. The technology is still almost wholly on the drawing board. Two of its main proponents, Ralph Merkle and Eric Drexler, worked with us at Xerox PARC. They built powerful nano-CAD tools and then ran simulations of the resulting designs. The simulations showed definitively that nano devices are theoretically feasible. But theoretically feasible and practically feasible are two different things. And as yet, no-one has laid out in any detail a route from lab-based simulation to practical development.

So here the road ahead is unpredictable not because of an unexpected curve, but because the road itself still lacks a blueprint. In the absence of a plan, it's certainly important to ask the right questions. Can nanotechnology actually fulfill its great potential in tasks ranging from data storage to pollution control? And can it do such things without itself getting out of control? But no one should worry too much about the road's maintenance crew when the road itself has yet to be surveyed. If the lesson of genetic engineering means anything, however, even though useful nano-systems are probably decades away, planners would do well to consult and educate the public early on.

Worries about *robotics* suggest that here, too, the route has been added to our mapbooks long before the road itself has actually been built. Take for example the much-talked about "bots" — the software equivalent of robots, which search, communicate, negotiate, or act as agents on the Internet. They, it has been claimed, do many human tasks much better than humans and so indeed might come to replace us all. In fact, bots are useful because they are quite different from humans. They are good (and useful) for those tasks that humans do badly. They are often quite inept at tasks that humans do well — tasks that call for judgement, taste, discretion, initiative, or tacit understanding. So bots are probably better thought of as complementary systems, not rivals to humanity. Consequently, though they will undoubtedly get better at what they do, such development will not necessarily make bots more human. They are in effect being driven down a different road. Certainly, the possibility of a collision needs to be kept in mind. In particular, we need to know who will be held responsible when autonomous bots inadvertently cause collisions — as well they might. But we probably need not look for significant collisions around the next few bends.

Are more conventional robots — the villains of science fiction — any greater threat to society? We doubt it. PARC research on self-aware, reconfigurable polybots has pushed at new robotic frontiers. When these are combined with our MEMS (microelectical mechanical systems) research, they point the way to morphing robots whose ability to move and change shape will make them important for such things as search and rescue in conditions where humans cannot or dare not go. Nonetheless, for all their cutting-edge agility, these robots are a long way from making good free-form dancing partners. In particular, like all robots (but unlike good dancing partners), they lack true conversational skills. The chatty manner of C3-PO still lies well beyond machines. Indeed, what talking robots or computers do, though it may appear similar, is quite different from human talk. Talking machines travel routes designed specifically to avoid the full complexities of situated human language.

True, robots may still seem quite intelligent. Yet such intelligence is profoundly hampered by their inability to learn in any significant way. (This failing has apparently led Toyota, after heavy investment in robotics, to consider replacing robots with humans on many production lines.) And without learning, simple common sense will lie beyond robots for a long time to come. Indeed, despite years of startling advances and innumerable successes like the chess-playing Big Blue, computer science is still almost as far as it ever was from building a machine with the learning abilities, linguistic competence, common sense, or social skills of a five year old.

So, like bots, robots will no doubt become increasingly useful. But, as a result of tunnel design, they will probably also become increasingly frustrating to use. In that regard they may indeed seem anti-social. But they are unlikely to be anti-social in the way of science fiction fantasies, with robot armies exterminating human society. Indeed, the thing that handicaps robots most of all is their lack of a social existence. For it is our social existence as humans that shapes how we speak, learn, think, and develop common sense. All forms of artificial life (whether bugs or bots) will remain primarily a metaphor for — rather than a threat to — society at least until they manage to enter a debate, form a choir, take a class, survive a committee meeting, join a union, build a lab, pass a law, engineer a cartel, reach an agreement, or summon a constitutional convention. It is these critical social mechanisms that allow society to shape its future. It is through planned, collective action that society forestalls expected consequences (such as Y2K) and responds to unexpected events (such as epidemics).

One Small Step for Futurology, One Large Step for Mankind

Why does the threat of a cunning, replicating robot society look so close from one perspective, yet from another quite distant? The difference lies in the well-known tendency of futurologists to count "1,2,3 ... a million" or even infinity. Once the first step on a path is taken, it's very easy to assume that all subsequent steps are trivial. So, for example, the telephone had barely appeared before people were predicting videophones — yet we still don't have videophones on any large scale today. Several of the steps Joy asks us to take — from genetic engineering to a White Plague, from simulations to out-of-control nanotechnology, from replicating peptides to a "robot species" — are extremely large. And they are certainly not steps that will be taken on an open highway without potholes, diversions, regulations, controls, or traffic coming the other way.

One of the lessons of Joy's article, then, is that the path to the future can look simple (and sometimes simply terrifying) if you look at it through what we call 6-D lenses. We coined this phrase having so often come upon "de-" or "di-" words like *demassification, decentralization, disintermediation, despacialization, disaggregation,* and *demarketization* in futurology. These are grand forces which some futurists see technology blowing through society and uprooting our social systems like an irresistible storm. If you take any one of the Ds in isolation, it's easy to follow its relentless journey to a logical conclusion. So, for example, because firms are getting smaller, it's easy to assume that firms and other intermediaries are simply disintegrating into markets. And because communication is growing cheaper and more powerful, it's easy to believe in the "death of distance." But these Ds rarely work in such linear fashion. Other forces (indeed, even other Ds) are, we need to remember, at work. Some, for example, are driving firms into larger and larger mergers to take advantage of the social (rather than just technological) networks. Other forces are keeping us together despite the availability of great communications technology. So, for example, whether communications technology has killed distance or not, people curiously just can't stay away from the social hotbed of modern communications technology, Silicon Valley.

Importantly, the Ds do indicate that the old ties that bound communities, organizations, and institutions are being picked apart by technologies. A simple, linear reading then suggests that these will now simply fall apart. A more complex reading, taking into account the multiple forces at work, offers a different picture. Undoubtedly some communities, organizations, and institutions will disappear. But others will reconfigure themselves. So, while many nationally powerful corporations have shriveled to insignificance, some have transformed themselves into far more powerful transnational firms. And while some forms of community may be dying, others bolstered by technology are growing stronger.

Two hundred years ago, Thomas Malthus, assuming that human society and agricultural technology developed on separate paths, gloomily predicted that society was growing so fast, it would starve itself to death. A hundred years later, H.G. Wells similarly assumed that society and technology were developing independently. Wells, however, like many today, saw technology outstripping society. So he predicted that technology's relentless juggernaut would unfeelingly crush great swathes of society. Like Joy, both Malthus and Wells issued important warnings, alerting society to dangers it faced.

But by their actions, Malthus and Wells helped prevent the very future they were so certain would come about. These self-*un*fulfilling prophecies failed see that, once warned, society could wittingly and unwittingly galvanize itself into action. It could develop agricultural technology to increase the food supply dramatically. And it could develop social constraints to temper the exuberance of technology. Of course, this social action in the face of threats showed that Malthus and Wells were most at fault in their initial assumption. Social and technological systems do not develop independently. The two evolve together in complex feedback loops, wherein each drives, restrains, and accelerates change in the other.

Of course, once the social system is factored back into prognostication, the road ahead looks much more convoluted. It is difficult to know what might lie beyond the next bend and which way it is best to turn. But this much can certainly be said. Communities, organizations, and institutions are indeed the main brake by which society slows the destructive power of technology (and, indeed, accelerates its advantages). As new technologies emerge, old institutional forms (copyright and patent law, government agencies, business practices, social mores, and so forth) inevitably prove inadequate. Consequently, society has to develop new ones. Robert Putnam's new book, *Bowling Alone*, shows this process in action. The dawn of the 1900s brought unprecedented technological advances, including the introduction of cars, airplanes, telephones, radio, and domestic power. With these advances came first, unprecedented social disruption, and then a remarkable period of legal, government, business, and societal innovation — stretching from the introduction of anti-trust legislation to the creation of the American Bar Association, the Sierra Club, the American Red Cross, the NAACP, and the YWCA. Society, implicitly and explicitly, took stock of itself and its technologies and acted accordingly. The resulting social innovation has left marks quite as deep as those left by technological innovation.

To deal with recent unprecedented technological change and the disruption it may cause, we will need similarly extensive and unprecedented social reflection and similar organizational and institutional creativity. New social forces, however, take time to develop. And the more people ignore them, the more time development will need. But technological acceleration gives us ever less time. So first the public at large needs to become engaged in these debates and to understand that society and institutions are part of the whole picture — something technological tunnel vision obscures. That way we can all see where the brake lies. Then we need to consider how an educated public can help construct new social institutions. That way, we can start to apply the brake where necessary.

NATIONAL NEEDS DRIVERS FOR NANOTECHNOLOGY

*G. Yonas and S.T. Picraux, Sandia National Laboratories**

Abstract

Nanoscience and nanotechnology may turn out to have significant societal implications, as would be the case for any truly revolutionary advance in technology. We have identified three areas — natural resources, human condition, and security — where trends are raising significant social issues that will become drivers for technological change. To achieve a safe, secure world we must consider both global and national aspects of security, and the above issue areas are significant in this broader context. These problems are complex and require a life cycle systems approach for technological advances to contribute to real societal solutions. Finally, as with any radically new technology, the consequences of using nanotechnologies can harm as well as help mankind. It is up to society to debate and develop total and durable solutions.

Introduction

The Clinton Administration's National Nanotechnology Initiative was instituted to

> ...support long-term nanoscale research and development leading
> to potential breakthroughs in areas such as materials and
> manufacturing, nanoelectronics, medicine and healthcare,
> environment, energy, chemicals, biotechnology, agriculture,
> information technology, and national security. The effect of
> nanotechnology on the health, wealth, and lives of people could be
> at least as significant as the combined influences of
> microelectronics, medical imaging, computer-aided engineering,
> and man-made polymers developed in this century. (NSTC 2000)

We argue that a government research and development initiative of this scale should go one step farther: the breakthroughs sought should relate to the central emerging problems of our society. While curiosity and unforeseen discoveries will still motivate the science, the scientific effort should point in the general direction of contributing elements to systems solutions to the

---

* Sandia National Laboratories is a multiprogram laboratory operated by Sandia Corporation, a Lockheed Martin Company, for the United States Department of Energy under Contract DE-AC04-94AL85000

complex challenges that face our nation. We need to think at an early stage about how nanotechnology will affect "the health, wealth, and lives of people."

Coming from a national security laboratory, we tend to think of most of the potential nanotechnology applications as having national security implications. Figure 6.5 suggests that national security cannot be independent of global security. But global security encompasses many more dimensions than just the military. The consequences of economic and informational globalization, combined with emerging demographic changes, will bring new kinds of threats to national and international security. Individual national security in a world of global collapse will not be tenable. We will have to seek national security in a context of global security (upper left quadrant of Figure 6.5).

Societal / Security Implications

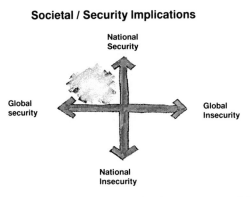

Figure 6.5. A Secure nation in a secure world.

At our nuclear weapons laboratory, we tend to think of problems in terms of systems and life cycles. (Sandia has responsibility for the non-nuclear systems in nuclear weapons from concept to production, to maintenance, and finally to dismantlement.) An example of a technology area where we as a nation did not work the entire life cycle problem is that of nuclear power. By not solving the nuclear waste disposal problem adequately as we developed the power generating systems, we left ourselves with a sizeable unresolved societal problem.

Our analysis of the conditions for global and national security leads us to consider three broad issues. The first revolves around the condition of the planet and its natural resources. We'll refer to this as the "green" issue involving foremost water, energy, and the environment. The second broad issue, which we call "red," is that of the human condition, with health at its center. These first two areas are potential sources of conflict that can drive global insecurity if unresolved. If it fulfills its promise, nanotechnology can enable solutions to many problems within these "red" and "green" issues areas. The third issue — which we call "black" — is military, for example as in the area of bio-warfare. Military advances enabled by nanotechnology, if used wisely in the interests of global security, can help to maintain a just peace. If used for purposes of aggression and domination, they can pose a substantial risk to all.

Natural Resources

There appears to be an increased potential for conflict as a rapidly growing world population tries to sustain itself with limited natural resources (Nichiporuk 2000; Brown, Flavin, and French 2000). The disparity in wealth between developed and developing nations, in combination with the uneven distribution of natural resources, remains a threat to the stability of states and of the international system. With the advent of modern manufacturing, advanced technologies, and the information age, the importance of natural resources has been reduced for developed nations. But, especially for developing nations, the availability and control of critical resources such as oil, water, and food on an increasingly crowded planet remain among the major sources of long-term insecurity. It is possible that nanotechnology will contribute to easing resource disparities. Potential areas of impact include new materials, potable water, new energy sources, and sustainable environmental processes.

The availability of water resources remains one of the big issues for potential insecurity around the globe. As the World Commission on Water for the 21st Century has pointed out,

> What is obvious is that progress, especially in developing
> countries, is much too slow, and that unless there are drastic
> changes, water shortages and environmental degradation will
> become the norm. More people than ever will be added to some of
> the areas of the planet that are already most vulnerable socially,
> economically, and environmentally. (World Water Council 2000)

Low cost techniques for water purification, self-cleaning, evaporation reduction, and desalination could have tremendous impact by providing adequate supplies of clean water. A major driver for regional conflict might be removed. Adequate water supplies are necessary not only for human health, but also to assure the availability of food for the developing world's growing population. The potential impact of nanotechnology on water supplies is hard to predict at this time, but several areas of significant opportunity come to mind. Affordable, engineered membranes that incorporated a self-cleaning process to avoid fouling could be used for large-scale desalination, which would go far in solving the water resource problem. While this technology would be a significant leap from current capabilities, the ability to tailor nanoscale membranes in combination with advances in self-assembly processes make it one to watch. In a variation on this concept, the ability to create membranes with molecular receptors that preferentially extract heavy metals and other pollutants is making progress in Department of Energy and other research laboratories (Roco et al. 1999). Another potential means of preserving water resources, particularly for agriculture, may be the control of evaporation through large-scale application of nano-engineered films or membranes. Management of water resources is a good example of where the life cycle systems approach should be taken to assure that the technologies employed do not leave unanticipated environmental problems in their wake.

A second "green" issue of growing long-term concern for global security is that of energy resources and their use. Although proven reserves of oil and natural gas are large, heavy energy usage by the developed countries, combined with the demographic and development trends of the third world, will eventually put pressure on the supplies. (With less than 5% of the world's population, the United States accounts for about 25% of world energy consumption.) In the meantime, the burning of fossil fuels has at least the possibility of substantially degrading the global environment.

Nanotechnology may be able to ameliorate energy problems both on the supply side and on the use side. In the near term, new, high-strength nanostructured magnets, nanolubricants, and other improved materials may greatly improve motor efficiency. In the long term, nanoengineered fuel cells, biocatalysts for crops for food or biomass fuels, or nanostructured photovoltaic films may permit cheaper alternative energy sources. For example, if the efficiency of photovoltaics were improved by a factor of two from the 20 to the 40% range at comparable costs — something that is theoretically possible — the role of solar energy would grow substantially. Likewise, if the oceans could be used for growing biomass fuels or harvesting energy through nano-biotechnology advances, significant increases in global energy supplies would result.

Systems life cycle thinking is particularly important in addressing the energy issue because of the coupling of energy and the environment. For example, if artificially engineered plants that produce ready-to-use energy become possible, at an early stage we will have to address issues akin to those now arising from the field of genetically engineered foods. But, properly designed, systems using such technologies as photovoltaics, engineered photosynthesis, factory process heat re-use, or agricultural fuel production could lead to a world of sustainable energy, agriculture, and climate. Such an "open system biosphere" (see Figure 6.6) would clearly have enormous implications for global security.

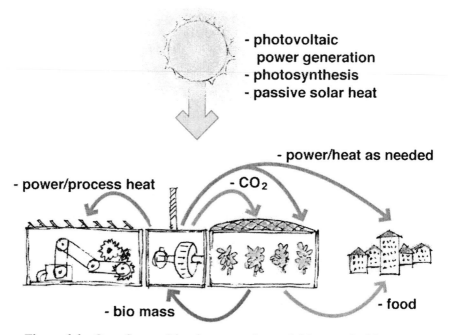

Figure 6.6. Open System Biosphere — a city model for sustainable energy, agriculture, and climate.

Nanoscience may also enable new materials and technologies that reduce economic dependence on other kinds of natural resources. The dependence of nations on extraction resources might be altered if common materials could achieve the functions of rarer and more costly materials. We refer to the ability to nanostructure a material for specific desired properties not found in its usual forms as *nano-alchemy*. In essence one is creating a new material by nanostructuring rather than by merely changing the chemistry. For example, common materials in the form of nanoscale clusters have been demonstrated to take on specific chemical catalytic properties, superior to those of more expensive catalysts. At this point we cannot predict whether nano-alchemy will apply broadly, or at what cost, but it is possible to envision large changes in how industries work. Then the relative wealth and power of nations could change, as could some of the contributing sources of international conflict.

Note, however, that there are no guarantees that the unregulated marketplace will assure a distribution of the benefits of nanotechnology that brings widespread prosperity and tranquillity. It is also possible to imagine the new technologies being used in ways that help the rich get richer and the poor get poorer. Given that nanoscience is being funded on a large scale from public resources, it behooves us to think on a national level about how its fruits can be directed to enhance national and international prosperity and security

Human Condition

The human condition — the "red" issue — also must be considered in an analysis of the potential societal implications of nanotechnology. In the United States the proportion of the population at retirement age is increasing and will continue to grow rapidly over the next several decades, with a corresponding decrease in the available fraction of workers in the society. This trend has been strong in the developed countries where the birth rate has declined significantly, leading to low or even negative population growth, while life expectancy has been increasing. With an aging population, an increasingly large fraction of national and personal resources is being spent on health care. Here, we will not discuss the additional, very serious health issues, such as AIDS and other emerging infectious diseases that burden developing countries. We would note, however, that if the applications discussed above of nanotechnology to securing clean water were to prove out, they could help with the disease problems of developing countries by improving sanitary conditions.

Desirable goals for an aging population include maintaining productivity longer, providing affordable health care, and deploying assistive technologies that maintain independence longer. Achieving these goals would greatly reduce the burden that an unhealthy, dependent older generation would place on younger citizens. The economic and social benefits to the nation would be great. We consider here just two possible connections to nanoscience and nanotechnology: assistive means to maintain physical independence and tools to support cognitive capability.

In the area of assistive devices, the ability to see (eye repair or hardware to replicate the eye function) and to maintain mobility (prostheses and sensor-based systems) could contribute significantly to maintaining productivity and physical independence. Advances in micro and nanotechnologies hold promise for contributing to a wide range of assistive solutions, from prosthetic limbs that adjust to the changes in the body, to more biocompatible implants, to artificial retinas or ears. Other opportunities lie in the area of neural prosthesis and the "spinal patch," a device envisioned to repair damage from spinal injuries.

In the area of cognition, revolutionary technical advances could have great impact on individual productivity and independence. We do not understand the workings of the brain well enough to predict with any confidence that assistive devices will actually work. However, rapid advances in the intersecting nano-, information science, and biological sciences seem to promise significant surprises. Possible results include devices that enhance learning, cognition, judgement and decision making. Devices that helped people with dementia — nearly a third of the population over 85 — could have great impact. At the same time concerns about the use of artificial or assisted cognition for social control must be addressed.

As with the potential benefits of technologies relating to natural resource use, those relating to human health and quality of life also could end up being available only to small segments of the world's population. Today we talk of the "digital divide"; tomorrow it may be the "nano divide." Only the right combinations of public policy (from whence a significant part of the initial investment in nanoscience is coming) and free enterprise will lead to maximizing the societal benefits of the new technologies.

Security

There is little doubt that nanoscience and nanotechnology will carry implications for the use of force for military and civilian security. Military and police organizations would highly value enhanced situational awareness in a world of ambiguity, confusion, and asymmetric threats. The implications of advances in computing speed, higher density memories, enhanced sensing and communication, and microsystems that, individually or in swarms, may contribute to situational awareness and control are obvious. Nanoscience will enhance all of these technologies. Implications of such advances range from distributed early warning, assessment, and response systems, to enhanced decision support systems. New non-lethal weapons may also emerge.

One area in which our understanding is rapidly growing is that of the emergent behavior of collective systems (see Figure 6.7). For example, researchers are beginning to appreciate how bees, with limited individual capabilities and simple rules of interaction, are collectively able to complete complex tasks, such as finding and harvesting nectar. Nanoscience, understanding of cognition, and microtechnologies may combine to give us small, smart devices that sense, think, act, and communicate as swarms.

Robotic swarms might play important roles in both security situations and natural disasters where direct human presence would be dangerous or ineffective.

Nano ⟹ microdevice ⟹ swarm system ⟹ collective emergent behavior

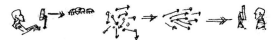

Figure 6.7. Nanotechnology may be a key enabling element to creating small, smart swarms of devices that sense, think, act, and communicate — resulting in emergent behavior of collective systems.

New information technology (possibly nanotech enabled) combined with better understanding of human and machine cognition, may give us new decision support systems. Information display and data fusion are already important military technologies. If memory aids (information storage and analysis) can be integrated with the human brain for decision support, applications in areas beyond military and emergency situations may become available. Related technologies would be interface devices such as wireless communication to the ear or displays on the retina, or reasoning support systems that would serve as decision advisors.

These various advances could contribute to global stability by enhancing the capabilities of peacekeepers to operate in difficult circumstances or of soldiers to resist aggression. As with other enhancements of military capability, however, they could also contribute to the success of military aggression. If the technologies were cheap and widely available, they could expand threats from terrorist or paramilitary groups.

Disruptive Technologies[1]

"Disruptive technologies" are those which produce new products in new ways. Initially, they may cost more and be less effective than the more mature, "sustaining technologies." But eventually, they become so much cheaper and better as to drive the older technologies out of the market. The technologies emerging from nanoscience may well prove disruptive. If so, they will have societal implications that extend beyond their functional

---

[1] A term coined by Clayton M. Christensen (Christensen 2000).

applications and into the realms of industry and economy (see Figure 6.8). Particular manufacturing firms, and perhaps entire industries (e.g. petroleum, agriculture), might be deeply changed, or even shrink to insignificance. Some managers and workers might be put out of business, while others may prosper. Those with the resources and adaptability to retrain may succeed, while others — perhaps especially older workers — may not make the transition successfully. Redistributions of economic power could lead to corresponding redistributions of political influence.

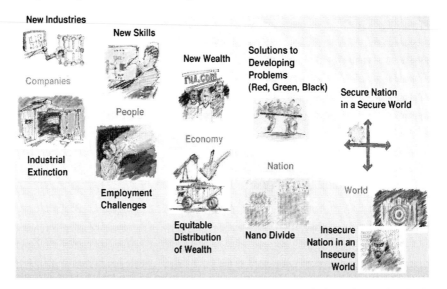

Figure 6.8. Nanotechnology may fall into the category of disruptive technologies where significant new capabilities and industrial systems bring large-scale changes, which may result in the betterment of society or may create new problems.

The international status of the nations which first master the new technologies may rise, while the nations overly committed to old industrial processes or to extracted resources may fall behind. As on the national level, redistributions of global technological strength could result in realignments of global prosperity and influence. These changes could promote national and international stability and security — or they could hinder it.

Conclusions

Nanoscience and nanotechnology may turn out to have significant societal implications, as would be the case for any truly revolutionary advance in technology (Figure 6.8). We have identified three areas — natural resources, human condition, and security — where trends are raising significant social issues that will become drivers for technological change. To achieve a safe, secure world we must consider both global and national aspects of security, and the above issue areas are significant in this broader context. These problems are complex and require a life cycle systems approach for technological advances to contribute to real societal solutions. Finally, as with any disruptive technology the advances brought about can be used for good or evil. It is up to society to debate and develop total and durable solutions.

References

Brown, Lester R., Christopher Flavin, and Hillary French. 2000. *State of the World 2000*. (Washington: Worldwatch Institute; see also http://www.worldwatch.org).

Christensen, Clayton M. 2000. *The Innovator's Dilemma: When New Technologies Cause Great Firms to Fail* (New York: Harperbusiness).

National Science and Technology Council (NSTC), Committee on Technology, Subcommittee on Nanoscale Science, Engineering and Technology. 2000. *National Nanotechnology Initiative: The Initiative and its Implementation Plan* (Washington, DC: Office of Science and Technology Policy, July)

Nichiporuk, Brian. 2000. "The Security Dynamics of Demographic Factors," RAND Corporation, MR-1088-WFHF/RF/DLPF/A.

Roco, M.C., R.S. Williams, and P. Alivisatos (eds). 1999. *Nanotechnology Research Directions: IWGN Workshop Report, Vision for Nanotechnology R&D in the Next Decade*. (Kluwer Academic Publishers; also available at http://www.nano.gov/).

World Water Council. 2000. *A Water Secure World: Vision for Water, Life, and the Environment* (Paris; see http://http://www.watervision.org/clients/wv/water.nsf/).

NANOTECHNOLOGY AND SOCIETAL TRANSFORMATION

M.M. Crow and D. Sarewitz, Columbia University

Remaking the World

Technological innovation sustains a fundamental tension of civilization — the tension between humanity's quest for more control over nature and the future, and our equally strong desire for stability and predictability in the present. Luddites were not against technology per se. They were against losing their jobs, and so they smashed the power looms that had put them out of work. The change wrought by technological advance continually remakes society, and this transformational process is on the one hand central to the dynamic that is commonly labeled "progress," and yet on the other is a source of continual destabilization and dislocation as experienced by individuals, communities, institutions, nations, and cultures.

In the age of science and technology (S&T), the federal government has increasingly become the prime catalyst for scientific advance and technological innovation. At the same time, modern government is also continually responding to and managing the transformational power of science and technology. Yet there is little effort to understand the relation between these two critical activities — advancing knowledge and innovation, and responding to their impacts — or to link them in a way that can enhance the value and capability of each.

A single technological innovation can remake the world. When the metal stirrup finally migrated from Asia to Western Europe in the 8th century, society was transformed to its very roots. For the first time, the energy of a galloping horse could be directly transmitted to the weapon held by the man in the saddle — a combat innovation of devastating impact. Because horses and tack were costly, they were possessed almost exclusively by landowners. Battlefield prowess and wealth were thus combined, and from this combination grew not just the traditions of a "warrior aristocracy" but the structure of European feudal society itself. Later, when the Anglo Saxon King Harold prepared to defend Britain against the invading Normans in 1066, he actually dispensed with his horse and ornamental wooden stirrups, choosing to lead his numerically superior forces on foot. The outnumbered Normans, however, boasted a strong, stirrup-equipped cavalry, and thus won the day — and the millennium (White 1962).

Such narrative has the ring of mythology, yet the experience of the industrialized world reinforces the knowledge that a new machine can help change everything. The invention of the cotton gin in the late 18th century allowed a vast expansion of cotton cultivation in the American south — and directly fueled a commensurate rise in the importation and use of slaves for plantation labor. One hundred and fifty years later, the mechanical cotton picker suddenly rendered obsolete the jobs of millions of African American share croppers, and catalyzed a 30-year migration of five million people out of the rural south and into the cities of the north. While the development of the mechanical cotton picker was no doubt inevitable, its proliferation was consciously accelerated by plantation owners who, fearing the rise of the civil rights movement, sought quickly to find a technological replacement

56

for the existing system of exploitation labor upon which they were economically dependent (Lemann 1991).

These examples point not only to the power of new technologies to transform society, but to the comprehensive interconnectedness of technological change and the complex social structure of society. The invention of the stirrup as a battlefield tool was in some very intricate way connected to the development and expansion of feudalism in Europe; the evolution of agricultural technology for a single cash crop is indissolubly bound to the ongoing struggle to overcome the U.S. legacy of slavery, segregation, and bigotry. More familiarly, a single class of technology — nuclear weapons — was a central determinant of geopolitical evolution after the end of World War II. Cars, television, air conditioning, and vaccinations have all stimulated foundational changes in society during the past century.

Of course new technologies rarely emerge in isolation. The industrial revolution is not just the story of harnessing steam power to factory production capability, but also the story of technological revolutions in transport, communication, construction, agriculture, resource extraction, and, of course, weapons development. These technological systems penetrated the innermost niches of society — home and family, school, workplace, community — and forced them to change. They also introduced completely new social phenomena, and stimulated the invention of completely new institutions.

The industrial revolution created the macroeconomic phenomenon of unemployment. Prior to the 19th century, even the most economically and politically advanced societies were dominantly agrarian and rural. For the majority of people, work was rooted in the home and the family. Vagaries of weather and transportation imposed irregularities and hardship, but most people and families harbored a diversity of skills that gave them independence from the marketplace and resilience to cope with a variety of challenges. In hard times, resort to subsistence farming and barter was usually possible (Keyssar 1986).

Industrialization and urbanization linked workers far more closely to the larger economic market, while removing the need and ability for them to maintain the diverse skills necessary for survival in the pre-industrial world. The traditional connection between manufacturing and agriculture in the home was sundered by new economic organization and by geography. Labor itself became a commodity, subject to the same fluctuations and influences as other commodities. During an economic downturn, factories fired people or closed down entirely. For the first time, workers could not easily respond to changing economic conditions by switching to a different type of work or moving to a subsistence mode. The political economist Karl Polanyi observed: "To separate labor from other activities of life and to subject it to the laws of the market was to annihilate all organic forms of existence and to replace them by a different type of organization, an atomistic and individualistic one" (Polanyi 1944, 163).

As technological innovation interacts with society to create new phenomena, such as unemployment, society also responds by developing new types of institutions and response mechanisms. Today we can recognize the problem of unemployment as central to a diversity of social, political, and economic structures and activities ranging from the organization of labor to insurance safety nets to educational programs to immigration policy. Unemployment rates are a key indicator of economic health, and a key determinant of political behavior. National and international economic policies focus strongly on managing unemployment, even as theoretical investigations seek to clarify the relation between unemployment rates and other key attributes of modern economies.

The general point is that transformational technology represents one variable in a complex assemblage of dynamic, interrelated societal activities. Decision making processes tend to address each of these activities in isolation from the others, e.g., conduct of research and development (R&D), dissemination of innovation products, development of regulations, reform of institutions. Concerted action occurs when a given innovation stimulates enough transformation to demand a response from other sectors of society. This response then triggers additional changes, which in turn demand further modulation. The process is reactive, discontinuous, disruptive, and sequential — like billiards. The challenge is to move toward a process of technology-supported societal progress where different sectors and activities can continually coevolve in response to knowledge about one another's needs and constraints — like an ecosystem. We are not there yet.

Transforming the Present

A brief consideration of evolution of information technologies helps to bring this look at societal transformation into the present. Gutenberg's perfection of the printing press of course had enormous transformational impact, allowing the broad dissemination of written texts and consequent expansion of information — and literacy — that undermined the Church's hegemony over knowledge and culture, and helped promote the dissolution of medieval social structure. Lewis Mumford suggested that the printed word represents "the media of reflective thought and deliberate action," a prerequisite, perhaps, for the intellectual achievements of the Enlightenment. But he also observed — as early as 1934 — that new modes of electronic communication were increasing the speed of information exchange to levels that made reflection impossible, and increasing the volume of information transmission to a point that exceeded our absorptive capacity (Mumford 1934, 240).

The implications of the information and communication revolution on democracy itself are far from clear. On the one hand, proliferation of information dissemination networks means greater access by more people to more information — and a greater capacity to communicate one's ideas and preferences in democratic fora. Control of information by authoritarian governments is becoming increasingly futile, and organization of democratic opposition increasingly enhanced, by new information technologies. But when this same capacity translates into 10,000 identical e-mail messages sent to a Member of Congress in support of a particular bill, one is hard-pressed to suggest that democracy is the beneficiary. Of particular concern is the recent increase in public referenda aimed at bypassing the legislative process. The barriers to putting referenda on ballots have been enormously reduced by information and communication technologies that can be used to disseminate ideas and organize group action with relatively little effort. While on the one hand this type of direct democracy can be a refreshing antidote to sclerotic legislative process, on the other it is quite often devoid of any serious deliberative process or public discourse, reflecting perhaps the pique of one well-organized interest group or individual, and the substantiation of a Warholian politics where anyone with access to a decent list-serve can lead a movement for a day. Is democracy in transition?

The implications of the information and communications revolution on the distribution of economic benefits in society are also problematic. Does the troubling increase in wealth concentration that characterizes both the U.S. and the global economy derive from the way that advanced technologies diffuse in market economies? Does the synergistic character of information and communication networks mean that disenfranchised populations and nations will find it increasingly difficult to participate in the spectacular economic growth that we have seen in the past decade? In other words, are the benefits of technology becoming increasingly appropriable by particular sectors of society, and is this in part an attribute embodied in new types of technological systems? Society is ill-prepared to answer such questions, let alone act on them in a knowledgeable manner.

Paradoxically, concerns about appropriability cut both ways. In the information society, the increasing ease of information dissemination may also threaten our system for protecting intellectual property and innovation. From pirated CD's sold on the streets of Shanghai to the advent (and apparent demise) of NAPSTER, the concept of intellectual property seems increasingly vulnerable. Are we looking to a future where such protection is no longer practically possible? Does a world without patents and copyrights seem unimaginable? More unimaginable than, say, the loss of monopoly over the written word would have seemed to the Church in 1450?

At issue here is not the value of change, but the path that change follows. What may look in retrospect like the march of progress may be experienced in real time as wrenching dislocation. The Dickensian squalor of 19th century London remains a symbol of the human impacts of technological change. Faced with unprecedented societal transformations, the English government (as well as other European states) failed to develop effective policies that could accommodate the rapid transition from rural agrarian to urban industrial society. Today, the plight of many overpopulated developing nations is the post-industrial, global manifestation of the same failure.

We see the fingerprints of societally transforming technological systems in the controversy over genetically modified organisms; in the morally reprehensible situation where 24 million HIV-positive sub-Saharan Africans cannot possibly afford AIDS drugs that are widely available in the affluent world; in the existence of 40 million Americans with no medical insurance; in the general inability of our public school systems to create a citizenry able to take advantage of the opportunities of the knowledge economy; in the challenges presented by the aging of our population; in the rising atmospheric carbon dioxide levels that symbolize 150 years of industrial dynamism.

Even the unprecedented rise of civil and ethnic conflict throughout the world in the past decade can be plausibly connected to technological transformation. Approaching this phenomenon from entirely different directions, the political scientists Samuel Huntington and Benjamin Barber each conclude that advanced communication and information technologies have created new fora for expressing ethnic identity and pursuing and strengthening cultural solidarity. Virtual communities, for example, can act to maintain identity over great distance, while also more efficiently garnering resources to support the expression of cultural goals. As Barber observes: "Christian Fundamentalists [can] access Religion Forum on CompuServe Information Service while Muslims can surf the Internet until they find Mas'ood Cajee's Cybermuslim document." The result may be locally empowering and globally divisive (Barber 1996, 155-156; Huntington 1996).

Nanotechnology and Societal Transformation

The marriage of science and technology beginning in the latter part of the 19th century accelerated the process of innovation, and thus the process of societal transformation as well. If the industrial revolution played itself out in less than 200 years, the electronics revolution seems likely to have a working life of perhaps 75 years, while the biotechnology revolution, although hardly yet on its feet, is already prophesied to be supplanted by (or perhaps to morph into) the nanotechnology revolution in the first half of the new century. What type of transformations might this revolution have in store?

Our point here is not to predict the future of nanotechnology and its impacts — an impossible goal — but to illustrate the direction and scale of thinking that will be necessary if we are to successfully manage the interaction of new knowledge and innovation with society. Judging by the literature prepared by the government (NSTC 1999; NSTC 2000), as well as the work of futurists and other techno-pundits (e.g., Cetron and Davies 1997), the promise of nanotechnology to remake our world seems virtually infinite. So the first thing to say is that if — as is variously claimed — nanotechnology is going to revolutionize manufacturing, health care, travel, energy supply, food supply, and warfare, then it is going, as well, to transform labor and the workplace, the medical system, the transportation and power infrastructure, the agricultural enterprise, and the military. Each one of these technology-dependent sectors is operated by and for human beings, who act within institutions and cultures, according to particular regulations, norms, and heuristics, all of which may reflect decades or even centuries of evolution and tradition. Not one of them will be "revolutionized" without significant difficulty. The current chaos in our medical system is emblematic of this type of difficulty.

In the near term, the current state of knowledge may suggest that the first wave of useful nanotechnologies will lie in the area of detection and sensing. The capacity to detect, precisely identify, and perhaps isolate single molecules, viruses, or other complex, nanoscale structures has broad application in such areas as medical diagnosis, forensics, national defense, and environmental monitoring and control. The potential for direct benefits is obvious; how might this evolving capacity influence society?

61

When detection outpaces response capability — as it usually does — ethical and policy dilemmas inevitably arise. For example, it is already possible to identify genetic predisposition to certain diseases for which there are no known cures, or to diagnose congenital defects in fetuses for which the only cure is abortion. In the environmental realm, new technologies that detect pollutants at extremely low concentrations raise complex questions about risk thresholds and appropriate remediation standards. The presence of tiny amounts of toxic materials in groundwater may justifiably raise alarm among the public even if the health risk cannot be assessed, and the technological capacity for remediation does not exist. These types of dilemmas may be expected to accelerate and proliferate with the advance of nanodetection technologies.

Advances in sensing and detection may transform existing societal mechanisms and institutions that were designed to cope with uncertainty and incomplete or imprecise information. The insurance industry, for example, deals with incomplete knowledge about the health of specific individuals by spreading its risk among large populations. If there is no way to distinguish between someone who is going to suffer a potentially lethal middle-age heart attack, and someone who is going to live to 105, then they can both get health and life insurance. Society clearly gains from this arrangement: costs are broadly disseminated, and benefits are delivered to those who most need them.

Medical sensors that can, for example, "detect an array of medically relevant signals at high sensitivity and selectivity" (NSTC 2000, 45) promise to aid diagnosis and treatment of disease, but also to develop predictive health profiles of individuals. Today, health and life insurance companies often use pre-existing conditions as a basis for denying or restricting coverage. The advent of nanodetection capabilities will considerably expand the information that insurance companies will want to use in making decisions about coverage. The generation of new information might thus destabilize the risk-spreading approach that allows equitable delivery of social benefits to broad populations. How will society respond?

Nanotechnology offers a dizzying range of potential benefits for military application. Recent history suggests that some of the earliest applications of nanotechnology will come in the military realm, where specific needs are well articulated, and a customer — the Department of Defense — already exists. One area of desired nano-innovation lies in the "increased use of enhanced automation and robotics to offset reductions in military manpower, reduce risks to troops, and improve vehicle performance." (NSTC 2000, 20). How might progress in this realm interact with the current trend toward rising civilian casualties (in absolute terms and relative to military personnel) in armed conflict worldwide? As increased robotic capability is realized in warfare, will we enter an era when it is safer to be a soldier in wartime than a civilian?

Such considerations are simple extrapolations of current trends in technological innovation and societal transformation. More adventurous speculation is tempting but is perhaps best confined to science fiction novels. The question of public response to nano-innovation, however, should not be avoided, even at this early stage. The ongoing experience of public opposition to old technologies such as nuclear power, new technologies such as genetically modified foods, and prospective technologies such as stem cell therapies, needs to be viewed as integral to the relationship between innovation and societal transformation.

Three observations are particularly relevant here. First, the impact of rapid technological innovation on people's lives is usually not consensual. Second, in the short term at least, the social changes induced by new technologies usually create both winners and losers (where what is lost may range from a job to an entire community). Third, rapid technological change can threaten the social structure, economic stability, and spiritual meaning that people strive in their lives to achieve. As the nanotechnology revolution begins to unfold in all its promise and diversity, such issues are bound to express themselves. They should not be viewed as threats, or as manifestations of intellectual weakness or repugnant ideology. Rather, they need to be recognized as a central part of the human context for technological change.

Preparing for the Revolution

> Now nanotechnology had made nearly anything possible, and so
> the cultural role in deciding what should be done with it had
> become far more important than imagining what could be done
> with it. (Stephenson 1995)

When resources are allocated for R&D programs, the implications for complex societal transformation are not considered. The fundamental assumption underlying the allocation process is that all societal outcomes will be positive, and that technological cause will lead directly to a desired societal effect. The literature promoting the National Nanotechnology Initiative expresses this view. The current policy approach thus addresses two elements:

- Conduct of Science and Technology

- Products of Science and Technology

These elements reflect the internal workings of the R&D enterprise. The fact that societal outcomes are not a serious part of the framework seems to derive from two beliefs: (1) that the science and technology enterprise has to be granted autonomy to chose its own direction of advance and innovation; and (2) that because we cannot predict the future of science or technological innovation, we cannot prepare for it in advance. These are oft-articulated arguments, not straw men. Yet the first is contradicted by reality, and the second is irrelevant. The direction of science and technology is in fact dictated by an enormous number of constraints (only one of which is the nature of nature itself). And preparation for the future obviously does not require accurate prediction; rather, it requires a foundation of knowledge upon which to base action, a capacity to learn from experience, close attention to what is going on in the present, and healthy and resilient institutions that can effectively respond or adapt to change in a timely manner.

If we flip the current S&T policy approach on its head, and start by thinking about desired social outcomes, rather than desired inputs to the R&D enterprise (i.e., more money), where would we begin? We might identify several very general categories of outcomes that most people would agree are worth thinking about. For example:

- Social equity: the distribution of the benefits of science and technology.

- Social purpose: the actual goals of societal development that we want to pursue or advance.

- Economic and Social enterprises: the shape and make-up of the institutions at the interface between technology and the human experience.

How can consideration of these types of outcomes be integrated into the S&T policy framework? The years since World War II have seen a very gradual evolution in the effort to connect thinking about S&T to thinking about the outcomes of S&T in society. A science policy report issued by the Truman Administration, for example, mentioned in its first pages the need to prepare for both the positive and negative impacts of scientific and technological change (Steelman 1947, viii). The rise of the environmental movement in the late 1960s reflected a public demand that society devote more S&T resources to the achievement of desired social outcomes like clean air and water. The creation of the congressional Office of Technology Assessment reflected growing public concern about the need to understand the societal implications of technological choices. Over the past decade, federally funded programs on the human dimensions of global climate change, and the ethical, legal, and social implications of the human genome project and information technologies, have been supported as adjuncts to much, much larger core research agendas in the "hard" sciences. Yet S&T policy itself remains input-driven.

Concepts such as sustainability, and analytical tools such as human development indicators, provide conceptual frameworks for linking R&D to societal outcomes, and in fact imply that outcomes are to some degree implicit in the choices we make about R&D inputs. These types of insights point the way toward the next step: to implement an approach to R&D policy that addresses the complex interconnections between technological advance and societal response. Such an approach would need to integrate the pursuit of innovation with an evolving understanding of how innovation and society interact, and include mechanisms to feed this understanding back into the innovation process itself. (In a very specific way, the private sector does this as a matter of course, as it uses consumer input to continually refine and improve the next generation of products.)

If we wanted to be serious about preparing for the transformational power of a coming nanotechnology revolution, we would need first to get serious — at this very early stage — about developing knowledge and tools for more effectively connecting R&D inputs with desired societal outcomes. This in turn would require the creation of a dedicated intellectual, analytical, and institutional capability focused on understanding the dynamics of the science-society interface and feeding back into the evolving nanotechnology enterprise. Such a capability might include the following elements:

- *Analysis of past and current societal responses to transforming technologies.* A case history approach could be used to investigate the diverse avenues that society has followed in responding to a range of technological advances. Understanding the roles and relations between the media, academia, policy makers, institutions, and cultural factors could be the basis for assessing — and anticipating — the likely trajectories of technology-induced social change.

- *Comprehensive, real time assessment and monitoring of the nanoscience and nanotechnology enterprise.* At this relatively early stage, it should be feasible to build a database of important activities in nanotechnology, and then track the evolution of the enterprise over time, in terms of directions of research and innovation, resources used, public and private sector roles, publications and patents, marketed products, and other useful indicators. This type of information is essential to understanding potential impacts.

- *A science communication initiative, to foster dialogue among scientists, technologists, policy makers, the media, and the public.* Understanding, tracking, and enhancing the processes by which information about nanotechnology diffuses from the laboratory to the outside world is central to understanding the social transformation process as it occurs. Of equal importance is the need to understand and monitor how public attitudes and needs evolve, and how they reach back into the innovation system. Empirically grounded, research-based investigations on communication can be the basis for strategies to improve social choice in ways likely to secure favorable outcomes.

- *A constructive technology assessment process, with participants drawn from representatives of the R&D effort, the policy world, and the public.* Technology assessment is both a process for bringing together a range of relevant actors, and an evolving product that can inform and link the innovation and decision-making processes. Understanding the changing capabilities of both the nanotechnology enterprise and various sectors and institutions likely to be affected by the enterprise can contribute to a healthy policy making environment where innovation paths and social goals are compatible and mutually reinforcing.

Should nanoscience and nanotechnology yield even a small proportion of their anticipated advances, the impacts on society will be far-reaching and profound — "as socially transforming as the development of running water, electricity, antibiotics, and microelectronics" (NSTC 1999, 1). We can allow these transformations to surprise and overwhelm us, and perhaps even threaten the prospects for further progress. Or we can choose to be smart about preparing for, understanding, responding to, and even managing the coming changes, in order to enhance the benefits, and reduce the disruption and dislocation, that must accompany any revolution.

References

Barber, Benjamin. 1996. *Jihad vs. McWorld: How Globalism and Tribalism are Reshaping the World* (New York: Ballantine Books).

Cetron, M., and Davies, O. 1997. *Probable Tomorrows: How Science and Technology Will Transform Our Lives in the Next Twenty Years* (New York: St. Martin's Press).

Huntington, Samuel P. 1996. *The Clash of Civilizations and the Remaking of the World Order* (New York: Simon and Schuster).

Keyssar, Alexander. 1986. *Out of Work: The First Century of Unemployment in Massachusetts* (Cambridge: Cambridge University Press).

Lemann, Nicholas. 1991. *The Promised Land: The Great Black Migration and How it Changed America* (New York: Random House).

Mumford, Lewis. 1934. *Technics and Civilization* (New York: Harcourt Brace Jovanovich, Inc.).

National Science and Technology Council (NSTC). 1999. *Nanotechnology: Shaping the World Atom by Atom* (Washington, DC: September; available at http://www.nano.gov).

National Science and Technology Council (NSTC). 2000. *National Nanotechnology Initiative: Leading to the Next Industrial Revolution* (Supplement to the President's FY 2001 Budget) (Washington, DC: February; available at http://www.nano.gov).

Polanyi, Karl. 1944. *The Great Transformation: The Political and Economic Origins of Our Time* (Boston: Beacon Press, 1957; orig. ed. 1944).

Steelman, John R. 1947. *Science and Public Policy* (Washington, DC: US Government Printing Office).

Stephenson, Neal. 1995. *The Diamond Age, or, Young Lady's Illustrated Primer* (New York: Bantam Books).

White, Lynn T. 1962. *Medieval Technology and Social Change* (London: Oxford University Press).

6.2 FOCUS ON ECONOMIC AND POLITICAL IMPLICATIONS OF POTENTIAL TECHNOLOGY

IMPACT OF NANOTECHNOLOGY ON THE CHEMICAL AND AUTOMOTIVE INDUSTRIES

J.M. Garcés and M.C. Cornell, Dow Chemical

We do what we can. This simple aphorism — the wisdom of the average person — may be used to shape the subject of this essay, namely, what can we do with nanotechnology in the chemical and automotive industries? Before trying to answer the question, we will take a brief scientific and historical detour.

The relationship between the meter and the nanometer can be understood better in monetary terms. A dollar bill is to a billion dollars, as a nanometer is to a meter. There are 1000 million dollars in a billion, and there are 1000 one thousand dollar bills in a million dollars. Thus a nanometer is 1000x1000x1000 times smaller than a meter, just as a dollar is 1000x1000x1000 smaller than a billion dollars. Nanotechnology deals with objects having at least one dimension in the range of nanometers, typically from 1 to 100 nm.

Scientists, from the onset of modern scientific thought in Greece, struggled for centuries to learn about the sizes of atoms and molecules. In 1905, Albert Einstein, as part of his doctoral dissertation, calculated the size of sugar molecules to be close to 1 nm. He used experimental results for the diffusion coefficient of sugar molecules in water measured by Graham, in the Einstein diffusion equation and obtained 0.99 nm as the answer. This historical development put the sizes of atoms and molecules on solid theoretical ground. Recent developments in microscopy [such as transmission electron microscopy (TEM) and atomic force microscopy (AFM)] allow scientists today to see and manipulate particles of nanometer dimensions as easily as a child can manipulate with a needle grains of salt seen with a light microscope. The TEM and AFM microscopes are about one thousand times more powerful than the light microscope.

The transition from theory to practice, from calculating the size of very small objects to manipulating them at will, defines the onset of nanotechnology. The things that we can see and do when we deal with nanometer size objects are the subjects of nanotechnology.

We can now leverage nano-scale particles into high-performance polymers and ceramics to yield composite materials having unique combinations of desired properties. We can beneficially exploit these composites in useful applications that provide enhanced functionality and value. The automobile is one platform that is beginning to take advantage of nano-composites in diverse components and systems ranging from catalytic converters that more efficiently convert combustion byproducts to benign emissions, to economical light weight plastics and coatings that enhance fuel efficiency and vehicle durability.

Catalytic converters, found in most modern cars and in high efficiency gas burners and wood stoves, utilize nano-scale metal oxide ceramic coatings to efficiently present a high surface area of precious metals like platinum and palladium to exhaust gases. These coatings are essential to speed and complete the conversion of harmful emissions such as carbon monoxide, nitrogen oxides, unburned hydrocarbons, and soot into benign by-products such as nitrogen, water and carbon dioxide. This is a good example of the beneficial impact of nanotechnology on the environment and on the quality of life, especially in large cities.

Nano-particles derived from clays and related materials, strengthen and stiffen plastics. This property is being exploited by converting inexpensive, light plastics to engineered composites that have the unique combination of both stiffness and toughness. Such composites are being developed for use in automotive bumper systems, for exterior body panels, and for instrument panel structures and interior closeout panels. Nano-reinforcement enables the homogenization of plastic materials, which will facilitate recycling of plastics upon disposal of a vehicle. The stiffening effect is also useful in the processing of plastics by increasing melt strength, which in turn enables the molding of larger and/or thinner parts without distortion. Blow molding — an inexpensive plastic manufacturing process — will especially benefit, enabling the production of lighter and less costly fuel tanks, bumper systems, and seating and other automotive and household plastic components. The use of polymeric nanocomposites in automotive applications will help to improve the efficiency of vehicles in miles per gallon and to reduce the volume of byproducts discharged to the environment.

It is hard to imagine nanotechnology without an intensive participation of the scientists and manufacturing technologies of the chemical industry. The assembly of atoms and molecules into materials and substances that are useful to society are essential to the high quality of life enjoyed by modern civilization. Pharmaceuticals, plastics, electronics, textiles, food and many other things are the result of human ingenuity translated into useful and cost effective products by the work of persons skilled in the use of technology to convert raw materials found in nature into manufactured products. Nanotechnology can extend the quality and number of useful products made by industry because it can provide new building blocks and new tools to assemble them and to convert them into new products. The size reduction of electronic circuits and components in the last forty years has resulted in the creation of new industries and in dramatic changes in our life styles. This incredible impact of size reduction in the electronics industry illustrates the potential of nanotechnology to change the future of all industries and the quality of life.

The transition from the eye to the light microscope expanded our field of vision by about 1000-fold. Modern electronics circuits have components that are visible with the eye and the light microscope. We have another 1000-fold of magnification open to us from the light microscope to the TEM and the AFM. We can only imagine the potential that this 1000-fold factor will have in the evolution of electronic devices and in the creation of new technology — nanotechnology — in all industries.

Carbon nanotubes is a good example of the new generation of nano-materials to be produced by the chemical industry. They are far stronger than steel, but lighter, and conduct electricity like metals. New applications for these amazing nanotubes are being reported daily, in electronics, chemistry, optics, and biology. In the automotive and plastic industries they are being examined for their reinforcing capability, but also to impart degrees of electrical conductivity to plastic composites. The utility of these features range from more efficient painting to potentially "plastic" circuitry integrated into insulating molded plastic articles. Static dissipative plastics efficiently accept paint, reducing volatile emissions in the application of solvent born coatings and enabling the use of non-solvent-containing powder coatings on plastics. They are also desired in fuel system components to prevent static discharge fuel ignition. Semi-conductive plastics are desired as functional enclosures to shield electronic components from disruptive electromagnetic interference, which is becoming more of a challenge as the automotive industry migrates to new electrical architecture.

Universities, new start-up and established companies, are engaged in the inventions needed to produce carbon nanotubes and other nano-scale materials and products at economic costs. They are conquering the barriers involved in working with very small objects. The rules and theories that govern these new forms of matter are the topic at the frontiers of science and technology. The challenges in dynamics, architecture, assembly, and fabrication of nanomaterials are reinvigorating all fronts of science: from engineering disciplines and physics, to synthetic organic and inorganic chemistries, from polymer and materials science and polymer fabrication, to electronics and biology.

Recent developments provide a glimpse of the new industries that will be created by the natural evolution of knowledge into technology. New mesoporous solids offer larger surface areas and larger pores available for catalysis, absorption and separations. These big brothers of the molecular sieves and zeolites hold promise for the design and development of new catalysts and devices able to hold larger molecules, and reaction intermediates, and to facilitate molecular events with the precision and elegance of natural enzymes. The marriage of micelles, owned by colloid chemists, with silicate chemistry, has created a new window that is taking scientists to explore the interactions of other macromolecules such as proteins and enzymes with all types of substrates. Combinatorial methods applied to bio-inorganic synthesis make it possible to find "wise" proteins and enzymes that grow crystals and materials replicating the elegance of coral growths, sea shells, or silicon chips. Thus, synthesis is moving from control of composition and structure to the control of size, shape, morphology and function. Soon we will have at our service the creative engines of nature to produce new materials with tools not very different to those used to make wine, cheese or beer.

Molecular electronic devices, such as redox switches, are proposed as components of new computer architectures to create chemically assembled electronic nano-computers from the "bottom up". Some of these devices are looking at DNA molecules as building blocks to fabricate circuits. This approach contrasts the "top down" technology used to manufacture electronic hardware today. Quantum dots, semiconductor-based nanoparticles, are luminescent materials that can be used in all sorts of devices based on optical signal detection. Coupling these nano-electronic and nano-optical devices will create a plethora of new technologies that will eventually displace the current "top down" built devices.

The chemical and automotive industries will be composers and musicians of the new harmonies to be produced by the evolution of nanotechnology. The creation of polymeric nanocomposites, new batteries, new electronic conductors, novel optical devices, catalysts and fuel cells, will lead to a transformation of the vehicle architectures into lighter, more efficient and high performance transportation products. The cost-performance balance will be a key driver of this metamorphosis from conventional to nano-structured systems and components. Society at large will be the beneficiary.

Examples of Nanotechnology Applications (from the report "National Nanotechnology Initiative: the Initiative and Its Implementation Plan," NSTC/NSET, July 2000)

a. Nanoparticle reinforced polymers

Requirements for increased fuel economy in motor vehicles demand the use of new, lightweight materials — typically plastics — that can replace metal. The best of these plastics are expensive and have not been adopted widely by U.S. vehicle manufacturers. Nanocomposites, a new class of materials under study internationally, consist of traditional polymers reinforced by nanometer-scale particles dispersed throughout. These reinforced polymers may present an economical solution to metal replacement.   In theory, the nanocomposite can be easily extruded or molded to near-final shape, provide stiffness and strength approaching that of metals, and reduce weight. Corrosion resistance, noise dampening, parts consolidation, and recyclability all would be improved.   However, producing nanocomposites requires the development of methods for dispersing the particles throughout the plastic, as well as means to efficiently manufacture parts from such composites.

Dow Chemical Company and Magna International of America (in Troy, MI) have a joint Advanced Technology Program (ATP) sponsored by the National Institute of Standards and Technology (NIST) to develop practical synthesis and manufacturing technologies to enable the use of new high-performance, low-weight "nanocomposite" materials in automobiles (NIST/ATP Project Number 97-02-0047).[2] The weight reduction from proposed potential applications would save 15 billion liters of gasoline over the life of one year's production of vehicles by the American automotive industry and thereby reduce carbon dioxide emissions by more than 5 billion kilograms. These materials are also likely to find use in non-automotive applications such as pipes and fittings for the building and construction industry; refrigerator liners; business, medical, and consumer equipment housings; recreational vehicles; and appliances.

b. Nanostructured catalysts

Researchers at Mobil Oil Co. have revolutionized hydrocarbon catalysis by the development of innovative nanostructured crystalline materials. Their program focused on zeolites, porous materials with well-defined shapes, surface chemistry and pore sizes smaller than 1 nanometer. A new zeolite

---

[2] See http://jazz.nist.gov/atpcf/prjbriefs/prjbrief.cfm?ProjectNumber=97-02-0047.

class, ZSM-5, was discovered in the late 1960s. ZSM-5 has a 10 atom ring structure that contributes pore sizes in the range 0.45 – 0.6 nm (smaller than in zeolites X, Y and larger than in A) and enables shape selected chemistries not previously available.

Zeolite catalysts now are used to process over 7 billion barrels of petroleum and chemicals annually. New Zealand is using the same catalyst to produce 1/3 of its oil fuel requirement by converting it from natural gas via methanol and then high-octane fuels. ZSM-5, along with zeolite Y, now provide the basis for hydrocarbon cracking and reforming processes with a commercial value that exceeds $30 billion in 1999 (J. Wise, Vice President Exxon, ret.). Another example at Mobil Oil Co. is the aluminosilicate 10 nm shaped cylindrical pores, which have been applied in both catalysis and filtration of fine dispersants in the environment (Liu and Mou, 1996). Further systematic advances in nanotechnology are expected to increase its share of an overall world catalyst market that exceeded $210 billion in 1999.

c. Amorphous metals with controlled atomic structure

Increasing ability to design and fabricate materials atom by atom has allowed creation of new materials with customized physical and electronic properties. An example of such a material is the amorphous alloy called Vitreloy™ ($Zr_{41.2}Ti_{13.8}Cu_{12.5}Ni_{10.0}Be_{22.5}$). The new material is twice as hard, twice as elastic, twice as strong, and twice as tough, compared to steel. The rebounding and heat transfer properties are significantly different from the crystalline materials due to the different types of atoms and their arrangements. The atoms in the amorphous alloy Vitreloy™ are in a densely packed, but random arrangement. Amorphous materials are formed by cooling the liquid material quickly enough to prevent crystallization; the atoms do not have time to arrange themselves into an ordered structure. Because of the varying sizes of these atoms and their random arrangement in the solid, there are no groups of atoms that can easily move past one another. A consequence of this low atomic mobility is the low internal friction when a force is applied.

Vitreloy™, discovered at the California Institute of Technology by W.L. Johnson in 1993, can be cooled from the liquid state at rates as low as 1°C/s and still form a completely amorphous solid. This slow cooling rate is very unusual for amorphous metal systems that often need to be cooled at far faster rates in order to prevent crystalline phases from forming. The unique properties of amorphous solids make them useful in many commercial applications. One of the first applications of Vitreloy™ has been in the design of golf clubs. The amorphous alloy is two to three times stronger than many other conventional materials like titanium and steel. Other applications include projectiles to alter the structure of subterranean oil fields and different defense equipment.

INFORMATION TECHNOLOGY BASED ON A MATURE NANOTECHNOLOGY: SOME SOCIETAL IMPLICATIONS

Thomas N. Theis, IBM T.J. Watson Research Center

Summary

The history of information technology has been one of learning to make "bits" smaller. There is no obvious and hard physical limit to the minimum size of logical devices that process information or the marks that store information. Indeed, quantum physics is being recast as a theory of information, and even a single atom can no longer be seen as the ultimate limit to the minimum size of a bit. Currently, the smallest logical devices being manufactured contain billions of atoms, and the smallest magnetic bits on commercial hard drives contain millions of atoms. Assuming continued exponential improvement in our ability to pattern matter at ever-smaller dimensions, in perhaps 35 years we will have the capability to design and control the structure of an object on all length scales from the atomic to the macroscopic — in other words, the beginnings of a mature nanotechnology.

To grasp some implications of a mature nanotechnology, imagine a world where information technology is truly ubiquitous and dirt cheap, where even trivial human artifacts contain extraordinary complexity and therefore extraordinary ability to process and communicate information. These broad capabilities of future information technology are easy to forecast, but their implications for society are still difficult to discern. History suggests that the most important future applications of the technology will surprise us. Rather than try to predict outcomes, I suggest some issues that society may struggle to resolve: current societal debates may provide some guidance for the future.

Our present quandary over copyright law may seem quaint in a future where reproductions of any object are increasingly inexpensive and increasingly indistinguishable from the original. Privacy issues will be ever more important in a world where ever more of the objects around us share information with each other. Finally, while cost does not appear to be a factor that will limit broad access to the benefits of information technology, education will be the key to full participation in the economy, even more than it is today. Stable resolution of these issues will take decades, but I am optimistic about the outcome

What is Nanotechnology?

In order to have meaningful discourse on the societal impact of nanotechnology, we must first agree on what we mean by nanotechnology. In fact, definitions vary. This is to be expected, since there is no agreed-upon educational curriculum for someone who wishes to become a "nanotechnologist." No university offers an advanced degree in nanotechnology, although faculty are thinking hard about it. Nevertheless, a growing community of researchers are beginning to call themselves nanotechnologists. Members of this community come from such diverse disciplines as condensed matter physics, synthetic chemistry, materials science, biochemistry, and electrical and mechanical engineering.

Apparently we are witnessing the emergence of a new technical discipline. This new discipline will benefit from and contribute to important advances in science, but I believe it will be primarily an *engineering* discipline. Several contributors to this forum have remarked on the cross-disciplinary nature of nanotechnology. This is a characteristic of emerging engineering disciplines, which synthesize principles and techniques from diverse scientific disciplines in the pursuit of building something.

But what are nanotechnologists going to build? Common answers are "things at the nanometer scale" or "things at the atomic scale". But that is not a complete answer. Chemists have been synthesizing ever more complex molecules for about two centuries. Materials scientists have been growing semiconductor crystals one atomic layer at a time for decades. To justify itself as a new engineering discipline, nanotechnology must be about more than the ability to build things with atom-scale precision.

In fact, nanotechnology is about the creation and manipulation of information. Since I work for a prominent information technology company, this statement may appear self-serving, but in fact, it stems from rather basic physics. Information is now understood to be a measurable, rigorously defined, fundamental construct of physics, on the same conceptual level as energy or entropy. Roughly speaking, the information content of a physical system is defined as the number of bits in its most concise description. (A bit is a "zero" or a "one" in the binary number system, thus a bit is the fundamental unit of information.) A perfect crystal has very little information content, since its structure can be described very concisely. All that is needed is a short string of bits to list the coordinates of silicon atoms to form a unit cell, and some bits to indicate repetition of the cell indefinitely to fill space. A perfect crystal is perfectly monotonous and therefore perfectly useless. But, if we supply the information needed to carve a particular pattern of impurities, metals and insulators into that crystal, it can become, for example, a microprocessor.

A typical microprocessor is one of the most complex, therefore information-rich, artifacts yet designed and built by our species. If we examine one closely, we find a nested hierarchy of structures. Some structures are as large as the entire silicon crystal (chip) and some structures consist of layers of differentiated materials only a few atoms thick. Integrating structure and function on many length scales down to the atomic, silicon microelectronics is one of the few existing examples of a true nanotechnology (see Figure 6.9).

Figure 6.9. A computer integrates structure and function on many length scales. A transistor includes some layers of differentiated materials only a few atoms thick. However, our ability to impart structure at scales less than 180 nm is limited. In contrast, biological systems are richly structured all the way to the atomic scale.

But if we measure all the structures in a microprocessor chip, we find the vast majority of dimensions to be on the order of 180 nanometers or greater.

180 nanometers happens to be the smallest principal dimension that can be routinely defined by the lithographic processes currently used in manufacturing. The smaller dimensions are achieved through controlled deposition of very thin films of material and a limited set of processing "tricks" which take advantage of forces at the atomic scale to help assemble the atoms. The amount of information we can incorporate into the structure is very limited at dimensions below 180 nanometers. Still, a microprocessor contains a lot more information than does a perfect crystal of silicon.

Now consider any living thing. Again we find a complex and nested hierarchy of structures, but now the hierarchy extends all the way to the sub-nanometer scale. Exchanging just two atoms in a molecule of DNA or RNA can make all the difference. A very long string of bits is required to describe even a single living cell. Indeed, the density of information in living things approaches the maximum that is physically possible. Furthermore, living things do not just contain information, but they continuously process vast amounts of information.

There is some correlation between the density of information (the complexity) in a structure and its ability to process additional information. That is basically why sometime next year my company and a few others will begin manufacturing microprocessors with 130 nanometer principal dimensions. Some years after that, we will manufacture at 100 nanometers, and so on. Thus, microelectronics research, development, and manufacturing will contribute to the development of nanotechnology.

But lithographic patterning techniques will never allow us to define complex hierarchical structures all the way down to the atomic scale. To do that, we must become much better at generating complex patterns the way nature does. We must take and extend the best from synthetic chemistry, condensed matter physics, materials science and biology, and learn to provide and control the precise conditions under which technologically useful structures will form through natural processes. Already we see exciting hints of what is possible from laboratories around the world. A recent example from IBM Research is shown in Figure 6.10.

None of this, by the way, involves self-replicating systems, the specter raised by Bill Joy in his recent article "Why the Future Doesn't Need Us" (Joy 2000). Natural assembly as currently pursued in our laboratories is kin to the processes through which water molecules assemble themselves to form a snowflake. A snowflake can have a complex hierarchical structure, but it is far from alive, and it does not replicate itself. Note, however, that when the right conditions are present, snowflakes are readily produced by the trillions.

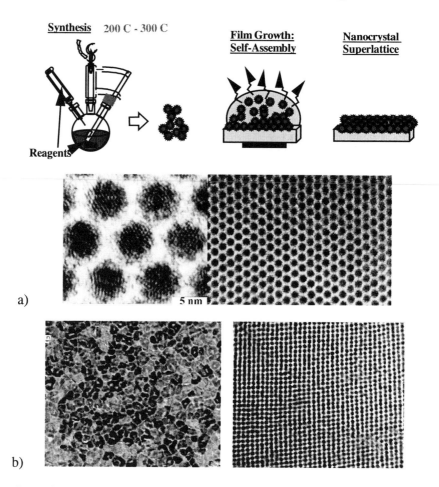

Figure 6.10. (a) Careful control and steering of process conditions can produce technologically useful structures on several length scales. Here chemical synthesis of uniformly sized nanocrystals and subsequent deposition of the particles from liquid solution forms a nanocrystal superlattice. (b) Comparison of current "state-of-the-art" magnetic storage medium with a nanostructured magnetic storage medium. Such nanostructured materials may contribute to cost-performance improvements in hard disk drives in the next few years (Sun et al. 2000).

This brings me to the last defining characteristic of a mature nanotechnology — very inexpensive manufacturing processes. The cost of microelectronics is currently dominated by the cost of the lithographic and related process tools used in manufacturing. Tooling costs increase dramatically every time the minimum lithographic dimension is shrunk. Processes of natural assembly can supplement and may someday eliminate lithography in much of our manufacturing. To be sure, the manufacture of complex, technologically useful structures will still require precise control of process conditions. However, we envision a class of manufacturing tools and processes which are simpler, more conservative of resources, and thus more cost-efficient than those of current practice. As we learn to build structures that are information-rich down to the atomic scale, the cost of information technology should continue to drop by many orders of magnitude.

Of course, nanotechnology will have many applications outside of information technology as presently defined. It will certainly enable new medical procedures and yield amazing new materials as suggested by other contributors to this forum. These will be incredibly information-intensive medical procedures and materials that will make our present "smart materials" look dumb indeed. Nanotechnology is, in a deep physical sense, concerned with the creation and manipulation of information.

To summarize:

- Nanotechnology is an emerging engineering discipline.

- An important focus of nanotechnology research is, and will be, understanding and harnessing natural processes of complex pattern formation for purposes of manufacturing.

- A mature nanotechnology will allow design and control of the structure of an object on all length scales from the atomic to the macroscopic, and will allow the manufacture of such information-rich objects at low cost.

- Such objects will be able to store information at close to the maximum density and perhaps process information at close to the maximum efficiency allowed by classical physics.

How Nanotechnology Might Develop Over the Next Few Decades

The history of information technology has been a history of learning to make "bits" smaller. There is no obvious and hard physical limit to the minimum size of logical devices that process information or the marks that store information. Indeed, quantum physics is currently being recast as a theory of information, and even a single atom can no longer be seen as the ultimate limit to the minimum size of a bit. Moreover, the smallest logical devices being manufactured currently contain billions of atoms, and the smallest magnetic bits on commercial hard drives contain millions of atoms. From a scientist's point of view, there is a long way to go in the development of information technology. However, as observed by John Seely Brown and Paul Duguid in their essay elsewhere in this volume, "Don't Count Society

Out" (see p. 37), the road ahead is not straight! It is not straight even in a purely technical sense. At current rates of progress, logical devices based on silicon and information storage based on magnetism will reach physical limits to size reduction in perhaps ten to fifteen years. Unless entirely new logical devices and entirely new storage devices are invented and brought to manufacturing, important economic forces driving the development of nanotechnology will be weakened.

Another twist in the technical road will occur when lithographic manufacturing processes can no longer be extended to smaller dimensions and must be supplemented or replaced by processes of natural assembly as described above. Rather than carrying silicon microelectronics and magnetic storage relentlessly forward, such technologies may first be established in niche applications where the cost of the established technologies can be easily undercut. The manufactured structures may not be particularly small, as shown by the examples in Figure 6.11.

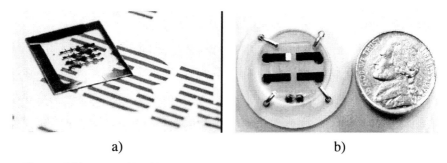

a)   b)

Figure 6.11. A world-wide community of researchers is currently pursuing the dream of printing or stamping electronic devices using organic molecules chosen to take advantage of natural assembly processes. (a) Transistors fabricated from the organic semiconductor pentacene exhibit performance comparable to the amorphous silicon transistors used in flat panel displays. (Dimitrakopoulos et al. 1999) (b) Yellow light from a diode made from a hybrid organic-inorganic compound, which was crystallized at room temperature from liquid solution (Chondroudis and Mitzi 1999).

But let us ignore for the moment the likely twists in the technical road, and instead assume continued exponentially compounding improvement in our ability to build complex structures with ever-smaller critical dimensions. Worldwide competition in microelectronics will certainly drive us in that direction for at least the next decade, but even at the end of a decade, atoms will be small. An additional 25 years will be required to develop the capability to fully design and control the structure of an object on all length scales from the atomic to the macroscopic — in other words, the beginnings of a mature nanotechnology. Society has some time to cope with, react to, and learn to harness this capability.

Some Potential Societal Implications of Nanotechnology

To grasp some implications of a mature nanotechnology, imagine a world where information technology is truly ubiquitous and dirt cheap, where even trivial human artifacts contain extraordinary complexity and therefore extraordinary ability to store, process and communicate information. These broad capabilities of future information technology are easy to forecast, but their implications for society are still difficult to discern. History suggests that the most important future applications will surprise us. Who, among the visionaries, architects, and early developers of the Internet, predicted electronic commerce as its "killer app"? History suggests that the trivial applications will also surprise us. What a leap of faith it would have been, if the builders of ENIAC (the first vacuum tube computer) could have imagined that in fifty years the computational equivalents of ENIAC would be built into greeting cards, for the sole purpose of saluting each card recipient with a trite melody, each "ENIAC" to be discarded after twenty seconds of playing time.

We would have to make similar leaps of faith in order to guess the impact of nanotechnology on society in fifty or one hundred years. I will not try. I limit my remarks to the next few decades. Rather than try to predict outcomes, I only suggest some issues that society may struggle to resolve. Furthermore, since a primary impact of nanotechnology will be to broaden and extend the reach of information technology, I suggest that many key issues are already foreshadowed by our present (still early!) experience with widespread dissemination of information technology.

Much public discourse currently centers on the ease with which copies of digital data can be made, what constitutes fair use of this capability, possible technical solutions to limit use, and the impact of all this on copyright and intellectual property law. Our present quandary over copyright law may seem quaint in a future where reproductions of any object are increasingly inexpensive and increasingly indistinguishable from the original. Bill Gurley (Gurley 2000) suggests that Napster, while a big part of this, is not the driver. Rather, the costs of digital information storage and communications bandwidth have dropped to the point where it is now feasible for a large population to copy, store, and transfer music digitally. Gurley points out what has just become feasible will rapidly become trivial. At current rates of compounding progress, sending a complete music CD to friend as an e-mail attachment will, in a few years, require no more relative storage capacity or bandwidth than sending a text file today. A typical home computer hard drive will easily house the entire CD canon (150,000 titles) in something like 12 years. Nanotechnology has the potential to support this compounding cost reduction for decades.

Thus the moral and legal issues surrounding the digital copying of music may be raised again in broader and broader contexts in the coming decades. Right now it is text, music, pictures, and the like that can be easily copied, but digital descriptions are, in principle, possible for any object. In practice, it will be a very long time, if ever, before digital descriptions of arbitrary objects are possible at the atomic scale. No matter. Less accurate but more compact digital descriptions will be good enough. An MP3 file "ripped" from a compact disk is a copy of a copy of an original sonic event. The CD format does not digitally encode all the information present at the original event, and for sake of compactness, the MP3 format discards additional information. But once the information is in digital form, regardless of format or fidelity to some original, it can be readily copied an arbitrary number of times with no further loss of fidelity. The fact that MP3 files are imperfect copies does not diminish the intensity of the societal debate regarding their proper use. In the coming decades, an increasing fraction of all the property in the world will be in the form of digital files or will be objects which, at ever diminishing cost or increasing resolution, can be "captured" as digital files. The instruments and manufacturing processes of nanotechnology imply the ability to reconstitute these digital files as "analog" objects when needed. Much good can result from this, and much mischief. It is very important that society get the rules right for the early test case of music.

Privacy and security in a networked world is another current topic of intense public debate. I suspect that this is another issue that will not be fully resolved for decades as nanotechnology allows the incorporation of information storage, processing, and communication in everyday objects at ever decreasing cost. Eventually, nearly every object around us might be networked. The current proliferation of web-enabled cell phones and hand-held computers is just the beginning. We glimpse the future as researchers strive to reduce the unit cost of radio frequency identification (RFID) tags to pennies. Eventually every manufactured object might keep a digital record of its manufacture, distribution, and use. The benefits have been widely discussed, but the potential for invasion of privacy is clear. Today we fear misuse of data aggregated from our transactions with various web servers. Tomorrow we may fear the misuse of data aggregated from our transactions with our clothing and household appliances.

Finally, as a society, we will continue to debate the best ways to ensure broad access to the benefits of information technology. Given what I have said about the likely low cost and broad dissemination of products based on nanotechnology, it appears that physical access can be assured to virtually everyone. But if nanotechnology becomes a dominant manufacturing technology, then manufacturing will be ever more concerned with information and dominated by information workers. Education, especially technical and scientific education, will be the key to full participation in the economy, even more than it is today.

Stable resolution of these issues will take decades, but I am optimistic about the outcome. In our pluralistic society, all viewpoints will be openly and passionately promoted. Current debates over fair use of information technology will generate the initial practices, policies, and laws that will help shape our nanofuture. Unforeseen applications of the technology and new societal issues will continue to arise. But there will be time, and in a democratic society there will certainly be opportunity, to make the necessary mid-course corrections.

Example of Nanotechnology Applications (from the report "National Nanotechnology Initiative: the Initiative and Its Implementation Plan," NSTC/NSET, July 2000): Giant magnetoresistance in magnetic storage applications

Within ten years from the fundamental discovery, the giant magnetoresistance (GMR) effect in nanostructured (one dimension) magnetic multilayers has demonstrated its utility in magnetic sensors for magnetic disk read heads, the key component in a $34 billion/year hard disk market in 1998. The new read head has extended magnetic disk information storage from 1 to ~20 Gbits/in^2. Because of this technology, most hard disk production is done by U.S.-based companies.

In 3 to 5 years, nonvolatile magnetic random access memory (MRAM) using the giant magnetoresistance phenomenon will be competed in the $100 billion RAM market. In-plane GMR promised 1Mbit memory chips in 1999. Not only has the size per bit been dramatically reduced, but the memory access time has dropped from milliseconds to 10 nanoseconds. The in-plane approach will likely provide 10-100 Mbit chips by 2002. Since the GMR effect resists radiation damage, these memories will be important to space and defense applications.

The in-plane GMR device performance (signal to noise) suffers as the device lateral dimensions get smaller than 1 micron. Government and industry are funding work on a vertical GMR device that gives larger signals as the device dimensions shrink. At 10 nanometer lateral size, these devices could provide signals in excess of 1 volt and memory densities of 10 Gbit on a chip, comparable to that stored on magnetic disks. If successful, this chip would eliminate the need for magneto-mechanical disk storage with its slow access time in msec, large size, weight and power requirements.

References

Brown, J.S. , and Paul Duguid. 2001. Don't count society out. *Societal Implications of Nanoscience and Nanotechnology.* National Science Foundation. Arlington, VA. March.

Chondroudis, K. and D.B. Mitzi. 1999. Electroluminescence from an organic-inorganic perovskite incorporating a quaterthiophene dye within lead halide perovskite layers. *Chemistry of Materials.* Vol 11. p. 3028.

Dimitrakopoulos, C.D., S. Purushothaman, J. Kymissis, A. Callegari, and J.M. Shaw. 1999. Low-voltage organic transistors on plastic comprising high dielectric constant gate insulators. *Science.* Vol. 283, 5 February. pp. 822 - 824.

Gurley, Bill. 2000. Digital music: the real law is Moore's law. *Fortune.* 2 October. page 268.

Joy, Bill. 2000. Why the future doesn't need us. *Wired.* April. pp. 238 - 262.

Sun, S., C.B. Murray, D. Weller, L. Folks, and A. Moser. 2000. Monodisperse fept nanoparticles and ferromagnetic fept nanocrystal superlattices. *Science.* Vol. 287, 17 March. pp. 1989 - 1992.

SOCIETAL IMPLICATIONS OF SCALING TO NANOELECTRONICS

R. Doering, Texas Instruments

Scaling to Nanoelectronics

The integrated circuit (IC) is a main "engine" of today's high-tech/high-productivity economy. During the past four decades, it has continued to revolutionize the ways in which we work, communicate, learn, and are entertained. And this revolution is not over! To the extent that we can continue to miniaturize ("scale") the components of integrated circuits, we will continue to provide greater functionality and performance at lower power and lower cost. The resulting societal impact of pervasive, affordable, IC-based electronics is summarized in Figure 6.12. In fact, most of the key IC improvement trends have been driven primarily by scaling to smaller features. These trends are listed in Figure 6.13 (SIA 1999).

* ## Economic Growth
    - Development of "High-Tech" Industries and Workforce
    - High-Productivity Economy

* ## Personal Equality/Opportunity
    - Global/Portable Communications
    - Personal Access to Vast Information On-Demand (News, Education, Entertainment)
    - Empowerment of Individuals to Process Information (Networked PCs and PDAs vs. Mainframe Computers)

Figure 6.12. Societal impact of IC "scaling" via pervasive, affordable electronics.

* **Functionality**   (e.g., eDRAM, eFlash, analog, RF)

* **Integration Level**   (e.g., components/chip – Moore's Law)

* **Compactness**   (e.g., components per area *or volume*)

* **Speed**   (e.g., microprocessor clock MHz)

* **Power**   (e.g., laptop or cell phone battery life)

* **Cost**   (e.g., cost/function – historically decreasing at >25% / year)

Figure 6.13. Improvement trends for ICs enabled by "Feature Scaling".

Present integrated-circuit technology spans both the "micro" and "nano" regimes. During IC fabrication, the constituent films are grown/deposited as thin as 2 nanometers and patterned (via optical lithography and plasma etching) into horizontal features as narrow as 100 nanometers (0.1 micrometers) (SIA 1999). Within the next decade, the pattern dimensions may approach 20 nanometers (SIA 1999, Wong et al. 1998). Scaling to this level should drive IC manufacturing cost per transistor well below a micro-cent even though cost will rise at the silicon-area level due to the increasing complexity of processing ever-smaller devices and interconnects, as shown in Figure 6.14 (Doering and Nishi 2000).

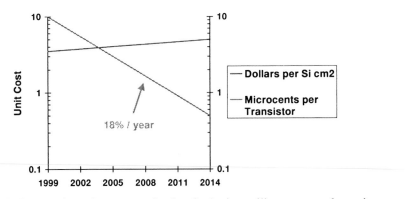

Figure 6.14. Manufacturing cost projection for logic at silicon-area and transistor levels

The resulting continued decrease in cost per electronic function should spur the creation of new business and consumer products and push annual worldwide sales of integrated circuits toward the trillion-dollar mark, as indicated in Figure 6.15.

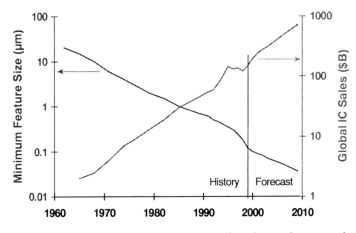

Figure 6.15. Smaller features → lower cost/function → larger market.

The new electronic products should also uphold their historic trend of "enabling/empowering the individual" — creating new opportunities for high-quality employment and for people to communicate with anyone, anywhere and to access and use (e.g., process) vast amounts of information on-demand. Nanotechnology can potentially help extend this vision even further into the future, via both new devices to serve as the "switches and interconnects" of the next IC era (Collier et al. 1999, Bockrath et al. 1997) as well as by providing a new manufacturing ("nanofabrication") paradigm. Note that some form of (perhaps "bio-based") nanofabrication might be very significant for cost reduction even if applied only to CMOS devices (Doering and Chatterjee 1998). These possibilities are outlined in Figure 6.16.

- In the future, how do we cost effectively manufacture "almost atomically-perfect" Nano-Electronic ICs ? (based on silicon, nanotubes, or whatever)

- Nano-Tool Arrays ?
  - Multiple-Tip AFM lithography ?

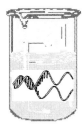

- Or Self-Assembly ?

  (perhaps DNA + Enzymes !)
  - Processing in a low-cost "beaker" ?
  - "Defect immunity" and/or self repair ?

Figure 6.16. Could "nano-fabrication" enable either: (a) continued scaling to new devices or (b) continued reduction of CMOS cost ?

If we are successful in developing nanoelectronic devices and nanofabrication techniques for manufacturing them in huge volumes at very low cost, the impact on society will be enormous. New businesses would emerge, many based on the capability to connect almost anything as a node in the "global network" (future version of the Internet). These businesses would help us identify, track, find, safeguard, inventory, control, diagnose, repair, upgrade, etc. virtually any type of item. The full scope of this vision represents a revolution in communication, with a large fraction of all man-made objects brought "on-line" via at least low-power wireless networking through nearby nodes. There would also, of course, be revolutions in computing based on nanoelectronics. At the high-end, enormously greater parallel processing with much more power at each node would greatly extend our ability to simulate complex systems, such as the weather, protein behavior, etc. On the portable side, computing at a level beyond today's "high-end desktop" would become available in form-factors limited only by human-interface convenience. For example, the visual display might be "heads-up in eyeglasses" and the input might be primarily voice. The non-I/O hardware would be negligible in size and weight. The entire system, including batteries and short-range wireless data/voice communications, could be integrated, for example, into "small-frame" eyeglasses. Note that the decreased power requirements enabled by nanoelectronics would be even more significant for this vision than the increased computing performance.

We suggest that the supporters of future research in nanotechnology should encourage broad cooperation between universities, industry, and government; emphasize precompetitive results; and include studies on technology choices/down-selection and technology migration/ displacement. These suggestions are listed in Figure 6.17.

- Start Early!
- Maximize Interdisciplinary Collaboration
- Use/Extend the ITRS as a Consensus-Building Forum on "Long-Range Research Needs" for Nanoelectronics
- Encourage Broad Cooperation
- Involve Industry as Stakeholders/Customers
- Utilize University Research Capability
- Leverage Federal/National Laboratories
- Emphasize Precompetitive Results (e.g., not proprietary)
- Include Studies on Technology Choices/Down-Selection and Technology Migration/Displacement

Figure 6.17. Suggested guiding principles for long-range nanotechnology research.

Societal Implications of Nanotechnology

The following potential general outcomes of research in nanotechnology should be of great benefit to society and especially to the electronics and information technology industry:

1. Continued rapid growth of the "high-tech" economy

2. Increased support of university research in the physical sciences, math, and engineering (reversing a long trend of decline)

3. Increased production of U.S. work-force educated in the above disciplines

4. Increased opportunity for collaboration between industry, university, and government researchers

5. Significant coordination of government and industry funding and review of academic research, especially through pre-competitive consortia

6. Increased synergy between historically separate research fields

Figure 6.18 illustrates the exponential growth of the worldwide semiconductor electronics market during the past four decades as more powerful and cost-effective integrated circuits have provided society with four "eras" of information technology. And, with the advent of worldwide networking, mobile wireless, and broadband capabilities, "information technology" now includes "communications" as well as "computing."

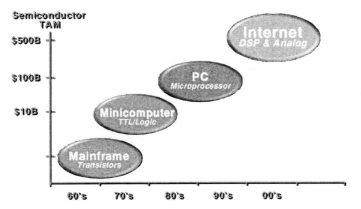

Figure 6.18. "Internet/Comm." is now driving IC demand to the next level.

There are also more specific potential results of research in nanotechnology that would have a profound impact on electronics and society. We have, perhaps, roughly another decade of rapid advance in integrated circuit technology based on the present paradigm of scaling CMOS. This is indicated in the 1999 International Technology Roadmap for Semiconductors (ITRS), which represents a consensus view on the future technology needs of the industry (SIA 1999). For example, Figure 6.19 shows some of the technology barriers (collectively known as the "red brick wall") that we are facing within the next decade.

At gate lengths of 20-30 nm, it is currently estimated that even dual-gate CMOS transistors may reach their practical scaling limits. One of the difficulties, threshold voltage "roll-off," is modeled in Figure 6.20 (Wong et al. 1998).

Year of Production:	1999	2002	2005		2008	2011	2014
DRAM Half-Pitch [nm]:	180	130	100		70	50	35
Overlay Accuracy [nm]:	65	45	35		25	20	15
MPU Gate Length [nm]:	140	85-90	65		45	30-32	20-22
CD Control [nm]:	14	9	6		4	3	2
T_{ox} (equivalent) [nm]:	1.9-2.5	1.5-1.9	1.0-1.5		0.8-1.2	0.6-0.8	0.5-0.6
Junction Depth [nm]:	42-70	25-43	20-33		16-26	11-19	8-13
Metal Cladding [nm]:	17	13	10		0	0	0
Inter-Metal Dielectric K:	3.5-4.0	2.7-3.5	1.6-2.2		1.5	<1.5	<1.5

Figure 6.19. Approaching a "red brick wall?" Challenges/opportunities for semiconductor R&D.

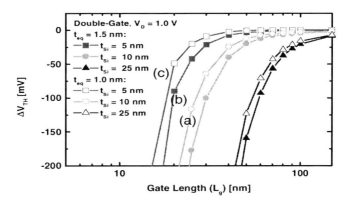

Figure 6.20. Is $L_g \cong 25$ nm the CMOS Limit? (Wong, Frank, and Solomon, IBM, 1998 IEDM).

Of course, improved system architectures and design techniques will extend improvement trends at the system level somewhat after the device and process technology mature. Beyond that point, corresponding to lithographic feature sizes approaching the 10-nm scale, fundamentally new types of "switches and interconnects" will probably be required to continue the basic "growth engine" underlying our "high-tech" economy. One of the great opportunities for nanotechnology is to supply such revolutionary electronic components, which would be needed by roughly 2015 at the historical rate of scaling (a component of "Moore's Law"). Carbon (and other) nanotubes (Mirsky 2000), quantum dots, and several potential "molecular switches" (Reed and Tour 2000) are all examples of current research that look very promising but need additional development to demonstrate feasibility for nanoelectronics. In addition to the development of new components, nanotechnology holds the promise of a fundamentally new manufacturing process, perhaps based on "biochemical techniques" (Doering and Chatterjee 1998). For nanoelectronics, this could represent the key enabler for continuing the historical decrease in cost per function available through integrated circuits (ICs). This in itself, even without new types of electronic components, could continue to fuel the high-tech productivity revolution for decades to come.

At this point, let's make a quick estimate of the potential impact of nanotechnology on the U.S. semiconductor industry via both of the aforementioned mechanisms. The 1999 annual sales of U.S. semiconductor makers was $77 billion, which increased by 4.1x (15% CAGR) as overall CMOS feature sizes were scaled down by 3.9x during the previous decade. Note that overall CMOS feature size is usually characterized in terms of "minimum half-pitch," which is also called the "technology node" (SIA 1999). Individual minimum features (isolated lines), particularly transistor gate lengths, were shrunk even faster during the past 10 years. The 1999 ITRS projections for scaling half-pitch and gate length over the next 15 years are shown in Fig. 6.19 (SIA 1999). If we assume that the 4.1x (~ 4x) sales increase was a result of IC cost reductions per function, performance increases, power reductions, etc. associated with the 3.9x (~ 4x) CMOS scaling, we could just approximate U.S. semiconductor company sales growth rate as equivalent to the rate at which we scale IC technology. Thus, we could estimate that sales would quadruple again, to about $300 billion, if we could scale CMOS from the 1999 state-of-the-art 180 nm technology node to 45 nm technology. Interpolation from Fig. 6.19 projects that this would be in 2012, which is also when transistor gate lengths would be getting into the aforementioned 20-30 nm range currently estimated to be the limit for CMOS devices. Thus, we might predict that U.S. semiconductor company sales growth would have to slow down from its historical 15% CAGR after it reaches $300-350 billion in another 10-15 years unless some form of device nanoelectronics is available to extend IC scaling beyond that period. This implies that the U.S. economy might stand to lose the difference between 15% and some "mature" growth rate (worst case, approaching just GNP growth) compounded annually on $300-350 billion after 2010-2015. Of course, as previously mentioned, nanotechnology in the form of a new low-cost manufacturing paradigm could also help to maintain the rapid growth of IC sales even if no new nanoelectronic devices were available to continue IC feature scaling. In fact, reducing cost-per-function is undoubtedly the most important single factor in the approximate proportionality between IC sales growth and feature scaling. Thus, we would expect to see most of the 15% growth rate extended beyond the $300-350 billion level even if we were still making CMOS at a fixed feature size, but with some form of "nano-manufacturing" which continued to provide significant reduction in cost-per-function (historically decreasing at about 25% annually).

The main risks for negative societal implications of nanotechnology will probably continue to be in the area of biotechnology rather than electronics. Traditionally, the largest societal risk associated with electronics has been in the area of system/component reliability. Of course, one of the great benefits of microelectronics has been its greatly improved reliability compared to older (e.g., vacuum tube) electronics. And it is expected that further miniaturization of electronics via nanotechnology should result in even further advances in reliability as nanoelectronic systems are designed and fabricated with atomic-level precision.

Of course, the positive impacts of nanoelectronics can also be expressed in more visionary terms. Continuing to improve the cost and performance of integrated circuits through nanotechnology would lead to a future in which almost everything that we manufacture and use could afford to include some electronic functionality. Even for the simplest items, this might include the ability to self-locate/inventory via low-power wireless communication through neighboring objects/nodes in the ubiquitous "global network." As one example, imagine how loss through misplacement and theft would be reduced if your car keys, watch, ring, etc. could send you a message with its current (GPS measured) location!

References

Bockrath, M., D.H. Cobden, P.L. McEuen, N.G. Chopra, A. Zettl, A. Thess, and R.E. Smalley. 1997. "Single-Electron Transport in Ropes of Carbon Nanotubes." *Science*, 275.

Collier, C.P., E.W. Wong, M. Belohradsky, F.M. Raymo, J.F. Stoddart, P.J. Kuekes, R.S. Williams, and J.R. Heath. 1999. "Electronically Configurable Molecular-Based Logic Gates." *Science*, 285.

Doering, R.R. and P.K. Chatterjee. 1998. "The Future of Microelectronics." *Proceedings of the IEEE*, 86, 1.

Doering, Robert and Yoshio Nishi. 2000. "Limits of Integrated-Circuit Manufacturing." *Proceedings of the IEEE.* September.

Mirsky, Steve. 2000. "Tantalizing Tubes." *Scientific American.* June.

Reed, Mark A. and James M. Tour. 2000. "Computing with Molecules." *Scientific American.* June..

SIA (Semiconductor Industry Association). 1999. "1999 International Technology Roadmap for Semiconductors."

Wong, H.-S.P., D.J. Frank, and P.M. Solomon. 1998. "Device Design Considerations for Double-Gate, Ground-Plane, and Single-Gated Ultra-Thin SOI MOSFETs at the 25-nm Channel-Length Generation." *Technical Digest of the 1998 International Electron Devices Meeting.*

FUTURE IMPLICATIONS OF NANOSCALE SCIENCE AND TECHNOLOGY: WIRED HUMANS, QUANTUM LEGOS, AND AN OCEAN OF INFORMATION

P. Chaudhari, IBM Watson Research Center

Why Nanoscale Science and Technology Now

If we use a length scale defined by a nanometer, nanoscience and technology have been around for several decades, particularly in research, development, and manufacturing in information technology, where film layers and lithographically defined features in the nanometer range are needed.

It is the wide availability of tools and information, initially produced primarily by the research community associated with information science and technology, to diverse scientific communities, outside of the information related communities, that has generated the current interest in this area. As a result, there is now a very significant broadening of the research base that is interested in nanoscience and technology. A notable example is the involvement of the biosciences community. It is this broadening that makes nanoscience and technology of interest now days.

Diffusion of knowledge, as epitomized by nanoscience and technology, from one discipline to another, so characteristic of all human endeavors, is a reminder that the interface between knowledge and ignorance migrates nonuniformly; rushing ahead in one area and then spreading laterally to others to produce an apparent broad uniform front, when measured over longer time scales.

My examples, concerning societal implications of nanoscience and technology, therefore focus on the future synthesis or symbiosis of the information and life sciences; nanoscience and technology is a vehicle for bringing different scientific communities together just as data mining or modeling by computers of the genome or protein folding brought the same two communities together, albeit with different emphasis.

It is a truism that we cannot predict the future. However, extrapolation of trends in science and technology and lessons from history can be used to guide us about the future. I want to use both to outline the future implications of nanoscience and technology in generating knowledge and new technologies, and of their potential impact on society. I shall do this by taking three examples to illustrate the kinds of profound changes that may arise. The description of these changes is perfunctory, as it must be, since I am opining.

Wired Humans

Computers have a very well defined historical trend. They have been reduced in physical size for a given capability, and their capabilities have continuously improved; these changes have been exponential with time. For example, a computer using vacuum tubes occupied space that was measured in tens of thousands of square feet. This was reduced to thousands of square feet with the introduction of the transistor, and to hundreds of square feet when the first integrated circuits were introduced. With miniaturization, the smallest commercially available computers have a footprint of some square feet to square inches. In addition, there is now a hierarchy of sizes and distributions, depending upon requirements. It seems to me, by simple extrapolation, that there is the possibility of further shrinking, to a point where computers can be carried by humans twenty-four hours a day.

Similar to, and parallel with, the evolution of computer technology, there has been an exponential change in communication systems. We have gone from fixed sites for initiating communication to the ubiquitous cell phone and personal digital assistant, which provide audio and video information. The networks, which collect information and carry it over distances before aggregation, have also become increasingly hierarchical. I am reminded of many dendritic or fractal structures in nature. Clearly, we are evolving to the point where every human being will be connected to any other human or to the vast network of information sources throughout the world by a communication system comprised of wireless and optical fiber communication links.

Humans communicate and receive information primarily using audio and visual means. Although tactile senses also provide information, I shall not discuss them here. Visual information requires large amounts of data to be transmitted and processed. It seems to me that in the long run this and the need for low-power devices will require the first base station to be not at a distance defined by a kilometer but a meter. In short, relay stations will be ubiquitous in the future. There is some evidence that this is already beginning — the bluetooth technology, for example.

Let us examine how these trends translate into some specific possibilities.

The human body is already wired by nature. Our nervous system operates on electrical impulses, generated by synaptic connections. There are many instances in which the electrical signals between the brain and an organ or a part of the body are disturbed or blocked. This could be due to an accident, disease, or defect; nanoscience and technology will play an increasingly important role here. Let me take an example to be concrete. Let us suppose there is a spinal injury, and signals to and from the brain can no longer be transmitted below the injured site. A chip can intercept the neuronal electrical signals, transmit them across the injured site, and then another chip couples the signal back to the body's nervous system. There can be many variants of this idea. There is already research in this area, but more is encouraged, for the benefits to mankind are obvious.

The next level of complexity involves nanoscale devices that can sense human temperature, pressure, or blood chemistry. These devices transmit, wirelessly, on a continuous basis the state of the human body. As long as the readings are within an acceptable level for a particular individual, no medical intervention is called for, but if they deviate from a defined range, intervention, by a human or a machine, becomes an option. There are already many ideas of using chips that can deliver drugs to local sites, and that bond, say, to a tumor and signal its whereabouts, and are wirelessly instructed to deliver a drug. Clearly, success of these possibilities can have profound implications for health care.

Let us carry these nascent ideas a bit further. Humans have an innate tendency to prefer not to carry devices or be bound by them. For example, the wireless telephone is preferred over one that needs a chord. It follows then, if one did not have to carry a personal digital assistant in one's hand and peer at a small display that would be even more desirable. If communication relay stations are all over, say within a meter range, then very low-power transmitting and receiving devices can be built. These can be carried or even implanted into adult humans. We know microphones can be implanted into throats, as is in the case of patients with cancer. Similarly, ear implants can be directly coupled to the mechanical sensors in the inner ear. More problematical, it seems to me, is visual information. Carrying large displays is a nuisance. Small projection displays have their drawbacks, similar to those of headphones. Perhaps a nanoscanning device, which directly projects a rastered image onto the fovea centralis, rather than the full retina may be invented. A ray of light will require compensation for refraction, if it has to traverse the cornea, but may not if it proceeds only through the vitreous medium. There is considerable room here for invention. If all of these speculations come to pass then a human will be wired fully — not only internally but also externally to the vast network outside of the body.

Quantum Science and Engineering: Quantum Legos

Our desire to understand nature has always been guided by a philosophy of reductionism. In order to comprehend the functioning of the very diverse world of nature, we invented the disciplines of physics, chemistry, materials, biology, etc. These partitions of nature have served us well. We are able to train students, develop methodologies, and perhaps most importantly, research subjects that interested us as individuals. The disciplines provided us an umbrella at academic institutions, within which we could advance our field, belong to, and be appreciated by. There was a similar evolution of disciplines in engineering but with a very different outlook. The engineers were not so much concerned with understanding the laws of nature but rather in using them to build something useful for mankind. In contrast to the reductionism of scientists, engineers are synthesizers.

Underlying most "hard" sciences are a handful of atoms. They combine in a myriad of ways to produce the world around us and including us. The physical scientists have worked hard to explain how things around us can be understood in terms of collection of atoms and molecules. Within the last decade or so, following the discovery of the scanning microscope and atom trapping devices, we have begun to "play" with single atoms or molecules; these have length scales at the low end of the nanometer and are quantized in their behavior. It is only a matter of time when these "quantum legos" will evolve into a (new) field of quantum science and engineering. Researchers working in this field-to-be are not likely to view their working philosophy as reductionism. They will synthesize new arrangement of atoms, learn ways to replicate them, and produce objects of interest to mankind. I can imagine that one day we will synthesize or build a molecule that, like the DNA, will store a code that leads to the production of material objects.

As in the case of nanoscience and technology, quantum science and engineering has also been around for a number of decades. It has not received wider recognition because it was also primarily confined to the information technology industry. It is used, for example, in the operation of transistors or lasers. Quantum mechanical considerations can be essential in many systems, where the number of atoms or the length scales exceed the nanometer range, for example, in superconductors or in photonic systems. Quantum cryptography, teleportation, or computers also fall into the broad field encompassed by quantum science and engineering.

An Ocean of Information: Thoughts on Privacy

I believe there will be numerous important contributions made by nanoscience and technology in many diverse areas of technology that benefit mankind. I have only touched on two of them. I believe it is only appropriate, as we are discussing the societal implications of this field, that we consider issues which go beyond science and technology. I want to raise this point, not so much because I have anything new to say, but rather in the spirit of raising issues that we must face and find solutions to.

Humans have always had the desire to generate, diffuse, and receive information about each other and the universe around them. What is changing very rapidly is the rate at which these three components are being implemented, the ease with which this can be done, and the nature of information. This virtual ocean of information about every aspect of life, including health, financial, and personal behavior, increasingly surrounds us. It has several thought provoking facets to it: our perception of reality and our sense of privacy are two examples. Here, I want to touch on privacy.

We often equate privacy with the fundamental human right of freedom. In the US, for example, the right to privacy is protected by the law (the law of torts), enshrined in the constitution (first, fourth and fifth amendments), and underpinned by a philosophy (Adam Smith) generally embraced by the people. It has deep roots. How will the people respond knowing that any information on an individual or a group can be fished out of this ocean? I do not know the answer but believe that the societal implications of any technology, which deeply touches, and to some sacred, social compact between people, deserves serious considerations.

IMPLICATIONS OF NANOTECHNOLOGY IN THE PHARMACEUTICS AND MEDICAL FIELDS

D.A. LaVan and R. Langer, MIT

Nanotechnology offers tremendous promise for advances in pharmaceutics and medicine. This revolution is transforming established disciplines such as biochemistry and enabling entirely new disciplines such as applied genomics. A distinction should be made between *molecular manufacturing*, the creation of molecules with highly specific shape and binding, and *solid-phase nanotechnology*, the creation of nanoscaled structures. Nanotechnology can be interpreted narrowly; these broader definitions are used for the purpose of this discussion. While some may dream of nanorobots circulating in the blood, the immediate applications in medicine will occur at the interfaces among molecular manufacturing, solid-phase nanotechnology, microelectronics, microelectromechanical systems (MEMS) and microopticalelectromechanical systems (MOEMS). Much work in these fields has been focused on developing new tools, techniques, and devices. The bounty will not be realized until those trained in these new paradigms begin to extend their research to address basic medical and scientific questions.

The post-genomics era has already begun to deliver on promises to provide detailed descriptions of cellular, molecular and genetic processes and pathways. While there are many years of work ahead to elucidate the human genome, discoveries are being announced regularly. Once a marker or receptor is identified, a molecule can be designed to interrupt abnormal cell behavior, stimulate the return of normal cell function, or specific binding can be used in a sensor to monitor normal biological cycles. These technologies will enable new diagnostic techniques, more specific therapies, and local delivery of drugs that will increase efficacy, slow the increase in resistance and reduce exposure to toxic compounds. Drews (1996) has reported that the number of drug targets resulting from the Human Genome Project is expected to be 3,000 to 10,000, compared with only 417 identified, empirically, to this point. In addition, the detailed understanding of the relationship between gene expression, molecular pathways and disease provides an opportunity to create highly specific, individualized treatments.

Combinatorial chemistry has begun to explore the new world revealed by the Human Genome Project. It is likely that the pharmaceutical industry will transition from a paradigm of "drug discovery" by screening compounds to the purposeful engineering of targeted molecules. Near term, current, approved, drugs can potentially be targeted to specific tissue by selective binding, improving the efficacy and reducing side effects (Arap, Pasqualini, and Ruoslahti 1998). It is important to recognize the node that occurs where molecular manufacturing and structural nanotechnology meet. For example, non-viral delivery systems for gene therapy may help to propel this industry; they conform to current regulatory models and their potential safety may more easily win physician and patient acceptance (Ledley 1995). Along related lines, Gref et al. (1994) entrapped up to 45 percent by weight of a drug in nanospheres, linked to polyethylene glycol derivatives, with extended circulation times due to decreased uptake by the mononuclear phagocyte system. Such systems may have value in altering drug biodistribution. Tobio et al. (1998) developed nanoparticles to deliver molecules, proteins and genes by transporting them through mucosal barriers. Putnam et al. (2001) and Lynn et al. (2000) have recently demonstrated polymers that condense plasmid DNA into nanostructures smaller than 150 nm with very little cytotoxicity in vitro[3].

An example of the coupling of microfabrication and nanotechnology is seen in the work of Santini et al. (1999), who demonstrated MEMS for delivering small quantities of a chemical substance on demand. This system could be coupled to sensors to fabricate an implantable pharmacy. The development of *in vitro* and *in vivo* diagnostic sensors will follow this path; they will have sensing elements produced by molecular manufacturing with power, telemetry and signal processing by a MEMS or MOEMS.

Current advances in diagnostic technology appear to be outpacing advances in new therapeutic agents. New molecularly based diagnostic techniques will become common, and traditional techniques will be improved. Highly detailed information from a patient will be available promoting a much more specific use of pharmaceuticals. For example, Piveteau et al. (2001) have recently succeeded, *in vitro,* in creating a targeted dendrimer by grafting galactose with nitroxide to enhance imaging contrast of the liver. In the near future, a healthcare provider may _easily_ identify genetic predisposition to a disease (Kalman and Lublin 1999), the virus or bacteria responsible for an infection (Gröndahl et al. 1999), or the health of a transplanted organ (Perkal et al. 1992). Many of these new tools will have a foundation in current techniques; a targeted molecule may simply add spatial or temporal resolution to an existing assay. Mahmood et al. (1999) have non-invasively imaged tumors (by adding spatial resolution) in mice by creating a near-infrared autoquenched fluorescent probe that targets proteolytic enzyme

---

[3] Measured only for the bulk polymer by Lynn et al. (2000)

activity. Automation of diagnosis may very well reduce the numbers of patients that require physician evaluation, reduce the time necessary to make a diagnosis, reduce human errors and enable wider access to healthcare facilities (i.e., via telemedicine).

It is early to predict the market impact that molecular manufacturing and solid-phase nanotechnology will have. A fair indication, because of the more immediate impact of these technologies on diagnostics, can be seen in the strong relative growth of the diagnostics sector recently in relation to the rest of the pharmaceutical industry. Total shipments in the diagnostic sector (all technologies) were estimated at almost 14 billion dollars for 2000. A plot of the data from Dun and Bradstreet is shown in Figure 6.21 (McConnell 1998).

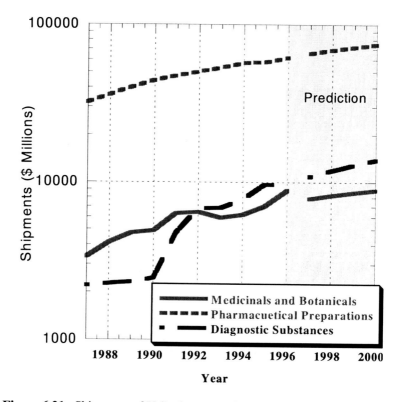

Figure 6.21. Shipments of U.S. pharmaceutical products by market sector (McConnell 1998).

However, there are obstacles for the widespread use of nanotechnology in pharmaceutical and medical applications. Public sentiment is rising against genetically modified food products, despite the great success and safe records of rDNA insulin, human-Growth Hormone, and hepatitis B vaccine. Every effort must be made to reduce risks to society, and the public must be educated to recognize the positive benefits of these new technologies.

One of the greatest obstacles to the development of highly specific drug therapies may result from *genetic discrimination* — discrimination directed against an individual or family based solely on an apparent or perceived genetic variation from the "normal" human genotype (Billings et al. 1992). Lapham et al. (1996) reported that perceived genetic discrimination caused some at-risk respondents to refuse genetic testing. The success of nanotechnology in certain areas of medicine (such as gene therapy) is dependent on successful policies to encourage patients to undergo genetic testing.

By using nanotechnology fundamental changes in drug production and delivery are expected to affect about half of the $380 billion worldwide drug production in the next decade. The U.S. company market share is about 40%. Nanotechnology will be used in various ways:

Example of Nanotechnology Applications (from the report "National Nanotechnology Initiative: the Initiative and Its Implementation Plan," NSTC/NSET, July 2000): Drug Delivery Systems

- Nanosizing will make possible the use of low solubility substances as drugs. This will approximately double the number of chemical substances available for pharmaceuticals (where particle size ranges from 100 to 200 nm).

- Dendrimer polymers have several properties (high solubility in aqueous solvent, defined structure, high monodispersity, low systemic toxicity) that make them attractive components of so-called nanobiological drug carrying devices.

- Targeting of tumors with nanoparticles in the range 50 to 100 nm. Larger particles cannot enter the tumor pores while nanoparticles can move easily into the tumor.

- Active targeting by adding ligands as target receptors on a nanoparticle surface. The receptors will recognize damaged tissue, attach to it and release a therapeutic drug.

- Increase the degree of localized drug retention by increasing the adhesion of finer particles on tissues.

- Nanosized markers will allow for cancer detection in the incipient phase when only a few cancer cells are present.

One example of current commercialization is liposome encapsulated drugs produced by Nexstar (doxarubicin for cancer treatment and amphotericin B for fungal infection) with sales over $20 million in 1999.

References

Arap W., R. Pasqualini, E. Ruoslahti. 1998. "Cancer Treatment by Targeted Drug Delivery to Tumor Vasculature in a Mouse Model", *Science,* Jan 16; 279 (5349): pp. 377-380.

Billings P.R., M.A. Kohn, M. de Cuevas, J. Beckwith, J.S. Alper, M.R. Natowicz. 1992. "Discrimination as a Consequence of Genetic Testing," *American Journal of Human Genetics*, Mar; 50 (3): pp. 476-82.

Drews, J. 1996. "Genomic Sciences and the Medicine of Tomorrow," *Nature Biotechnology*, Nov., 14 (11), pp. 1516-1518.

Gref, Ruxandra, Yoshiharu Minamitake, Maria Teresa Peracchia, Vladimir Trubetskoy, Vladimir Torchilin, Robert Langer. 1994. "Biodegradable Long Circulating Polymeric Nanospheres", *Science*, Vol. 263, 18 Mar, pp 1600-1603.

Gröndahl B., W. Puppe, A. Hoppe, I. Kühne, J.A. Weigl, H.J. Schmitt. 1999. "Rapid Identification of Nine Microorganisms Causing Acute Respiratory Tract Infections by Single-Tube Multiplex Reverse Transcription-PCR: Feasibility Study." *Journal of Clinical Microbiology*, Jan; 37 (1): pp. 1-7.

Kalman B., F.D. Lublin. 1999. "The Genetics of Multiple Sclerosis. A Review." *Biomedicine and Pharmacotherapy*, Sep; 53 (8): pp. 358-370.

Lapham, E.V., C. Kozma, J.O. Weiss. 1996. "Genetic Discrimination: Perspectives of Consumers." *Science*, Oct 25; 274 (5287): pp. 621-624.

Ledley, Fred D. 1995. "Nonviral Gene Therapy: The promise of Genes as Pharmaceutical Products," *Human Gene Therapy*, Vol. 6, Sep, pp 1129-1144.

Lynn, David M., Robert Langer. 2000. "Degradable Poly(β-amino esters): Synthesis, Characterization, and Self-Assembly with Plasmid DNA," *Journal of the American Chemical Society*, Vol. 122.

Mahmood U., C.H. Tung, A. Bogdanov Jr., R. Weissleder. 1999. "Near-Infrared Optical Imaging of Protease Activity for Tumor Detection." *Radiology*, Dec; 213 (3): pp. 866-870.

McConnell, S.A., ed. 1998. *Pharmaceuticals*. Dun and Bradstreet Industry Reference Handbooks, Gale, Detroit, pp. 30-34.

Perkal M., C. Marks, M.I. Lorber, W.H. Marks. 1992. "A three-year experience with serum anodal trypsinogen as a biochemical marker for rejection in pancreatic allografts. False positives, tissue biopsy, comparison with other markers, and diagnostic strategies." *Transplantation*, Feb; 53 (2): pp. 415-419.

Piveteau, Laurent-Dominique, Anthe S. Zandvliet and Robert Langer. 2001. "Nitroxide and Galactose Grafted Dendrimer: A Specifically Targeted Contrast Agent for MRI and EPR Imaging of the Liver," *in preparation*.

Putnam, D., Gentry, C.A., Pack, D.W., Langer, R. 2001. "Polymer-based Gene Delivery with Low Cytotoxicity by a Unique Balance of Side-chain Termini." *Proceedings of the National Academy of Sciences*, U.S.A., Volume 98, Number 3, pages 1200-1205.

Santini, John T., Jr., Michael J. Cima and Robert Langer. 1999. "A Controlled-release Microchip" *Nature*, Vol 397, 28 Jan, p 335-338.

Tobio, M., R. Gref, A. Sanchez, R. Langer and M.J. Alonso. 1998. "Stealth PLA-PEG Nanoparticles as Protein Carriers for Nasal Administration", *Pharmaceutical Research*, Vol. 15, No 2, pp 270-275.

WE'VE ONLY JUST BEGUN

R.S. Williams and P.J. Kuekes, Hewlett-Packard Labs

If anyone understands "future shock," it's those of us who live in and work in Silicon Valley. For the past decade or so, we've witnessed a constant preview of the future: much of what takes hold technologically throughout the world starts here.

But for all the change we've seen so far, in our view *the computer age hasn't even begun yet*.

A Look at the Past

Let us look at the primary technology that dominated the latter part of the previous century: the invention of the integrated circuit in 1959. Since that date, the number of transistors that can be fabricated onto a single chip has been doubling about every 18 months — a rate commonly known as Moore's Law. Moore's Law is an example of an exponential process: it has taken us from a crude chip with a single transistor to integrated circuits with 100 million active components in only 40 years.

At the same time, the amount of useful work that comes out of an integrated circuit for each unit of electrical power put into it has also increased by roughly 100 million. This astounding technological progress has given us what we call today the Information Age. It has had a profound effect on the lives and fortunes of people, companies and countries throughout the world.

How much longer can this exponential growth continue? In biological systems, the early stages of growth in any population are usually exponential. However, certain factors arise — such as limited resources, increased predation or a deteriorating environment — that can cause the process to slow. If you plotted the size of the population versus time, you'd get an "S" curve.

Ever since it was first proposed, there has been a great deal of discussion about when Moore's Law would reach a limiting plateau. We believe that the progress we've come to expect from silicon technology will reach its physical, engineering and economic limits in about 10 years. By that we mean that, not only will the never-ending quest for tinier and tinier transistors run into the limits of pure physics, the ability to manufacture them will also encounter a similar wall in the mechanical world. Finally, even if the physics and mechanics were attainable, no one company will be able to afford to make them. Says who? Says no less an authority than Gordon Moore himself, whose less well-known "Second Law" states that the cost of building factories also increases exponentially. Today's fabs cost about $3 billion to build. In 12 years, they may cost as much as $50 billion — a prohibitive amount of capital for any one company or even group of companies to raise.

The fact that silicon is approaching the flat part of its "S" curve, however, doesn't necessarily mean that progress in electronics or computing will slow. In fact, we believe the ultimate physical boundary for computing is a factor of 100 million beyond where we are today. That's why we say the computer age hasn't even begun yet.

What will take us well beyond 2010 are new technologies that are being pursued in university and corporate laboratories around the world, including our own. At HP Labs, in conjunction with our collaborators from UCLA, we are investigating the development of molecular or quantum-state switching devices and the design of systems that will assemble themselves through molecular recognition. These systems will be designed to operate perfectly, even if many of the components are defective, because we recognize that nature — in creating these self-assembling nanostructures — won't provide perfect parts. That's why computer scientists have joined our transdisciplinary team of physicists, chemists and engineers to design an architecture that will simply program around the defects. Incredible as all of this sounds, HP Labs computer scientists have already built an experimental supercomputer (using silicon technology) to test the concept of defect tolerance: it ran 100 times as fast as a workstation, despite having more than 200,000 defective components. Our UCLA collaborators have demonstrated that certain molecules can be utilized as electronic switches. Within the next year, we intend to build a memory using these molecules as an experimental proof of principle for molecular electronics. This memory will only hold sixteen bits of information, but it will fit in a square 100 nanometers on a side (for comparison, the smallest <u>wire</u> in a current generation Si chip is 180 nanometers wide).

Technologies — Bio, Info and Nano

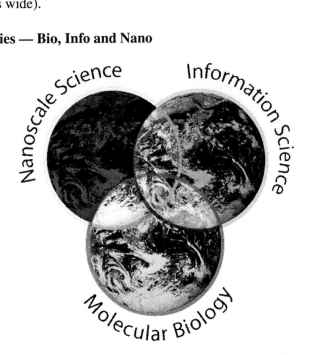

Figure 6.22. At the beginning of this millennium, we are watching the birth of three great new technologies: biotechnology, information technology and nanotechnology.

Our scenario for exponential growth during the next 20 years is not limited to electronic circuits. At the beginning of this millennium, we are watching the birth of three great new technologies: biotechnology, information technology and nanotechnology — we'll call them bio, info and nano for short. Bio is the rational utilization of the chemistry of life; info is the harvesting, storage and transmission of information; and nano is the control of all matter at the scale for which basic material properties are determined. All three of these areas will have completed the transition from applied science to technology during the next 20 years, and all three will see exponential growth in their capabilities. Each by itself would qualify as an "industrial revolution," but having all three progressing simultaneously — sometimes competing and often interacting with each other — will be completely beyond anything we have experienced to date.

How might the interaction of these technologies affect everyday life in 2020? Here are three scenarios. At first glance, none of these examples seems to be related to the web, but think of what happens when the Web itself is so pervasive that you literally forget it is there — it is the medium that carries information from anywhere to everywhere.

The first is "telepresence." We will have devices of sufficient sensory fidelity and information transfer capacity that, although they won't yet be perfect, they will give us the emotional experience of "being there." Not only will we be able to experience a sporting event from home as a fan in the stands would, but we could also face it from the perspective of any of the players or referees on the field. We could experience a drama from within the scene, either as a disembodied spirit or from the perspective of any of the characters. Scientists could have the "Fantastic Voyage" experience of traveling through the veins and arteries of a living being — not by shrinking themselves down, as depicted in the movie, but by receiving information from a remotely controlled sensor unit inside a subject that may be half a world away. Similarly, other explorers could fly over Mars, travel down the throat of an active volcano or stroll through a forest of molecules on the surface of a new catalyst.

The second is health care. As we now all know, it took about 10 years to sequence the first human genome. In 2020, getting a complete genetic map will be a standard test, much like a blood test is today. This will be a wonderful tool for doctors to assess risks and design specific treatments for genetically related conditions, but it won't help much for trauma victims or those who suffer from environmentally caused illness. For these cases, there will be three-dimensional whole body images with resolutions close to that of individual cells. A physical exam will include a complete head-to-toe

106

scan using several techniques simultaneously, with the data fused together not only to provide models of all internal organs, but also to label which cells are healthy and which are diseased. This will eliminate the need for exploratory surgery, and will guide minimally invasive procedures to remove tumors or repair damaged tissue.

The third is conversing with our machines. What will the appliances built in the next 20 years be able to do that they cannot do today? For one, they will be able to converse with people. By the year 2020, the data handling capacity of a "low end" electronic system will be roughly equivalent to that of the human brain. At that point in time, it should be possible for our electronic devices to pass a limited version of the "Turing Test" — in other words, to take part in a five-minute conversation with an individual so convincing that the person could not determine whether he was talking to another human being or a machine. This capability to converse in natural language should finally make the human-machine interface as natural as our interactions with other people. It will mean a significant shift in our view of the dividing line between what is natural and what is man-made, as well as produce entirely new goods and services that will seem essential then, but which we cannot even imagine today.

Economic and Social Consequences

What are the economic and societal consequences of these three exponentially advancing technologies? We will most likely see entire industries rise and fall within a period of a decade, as first one and then another of the new technologies addresses the problems that people face. This will create amazing opportunities for those who do not fear uncertainty and are willing to continually re-create themselves. But it will be extremely unsettling and disruptive for many, if not most, people. The only certainty will be change, and it will be extremely difficult for social and political systems to evolve quickly enough to keep up with technology. Moreover, because the technological progress will be exponential, the rate of change will also be continually increasing.

We may very well see the paradoxical situation that technology will — in absolute terms — improve the lives of nearly everyone on the planet, but most people may actually feel disenfranchised and less happy because the relative spread in wealth will continue to widen. In order to cope with increasingly rapid change, we have to force ourselves as a society to take the time to understand the consequences of our actions and ensure that our wisdom is increasing fast enough to keep up with the changes. We shouldn't fear technology itself, but we must beware the consequences of technological ignorance and irresponsibility. Technology does not emerge whole, like the fossil remains of a dinosaur uncovered by an archeologist. Rather it is a reflection of the decisions made by countless human beings, whether those decisions were informed or not.

The key is education — people must learn how to keep learning throughout their entire lives. Schools and universities may once again be thought of as institutions where students go to mature and stretch themselves intellectually, rather than to prepare for a specific career that may no longer exist by the time they graduate. The ability to adapt quickly to new environments, which has been the main differentiator of our species, will be challenged as a reaction to our technologies. And it's incumbent upon *all* of us — not just the technologists — but ordinary citizens as well as leaders in government, business, science and education to take a proactive part in shaping the use of technology for the good of our planet and the welfare of all. We really need to take the time to consider what is coming our way and how we should respond to it.

AN ECONOMIST'S APPROACH TO ANALYZING THE SOCIETAL IMPACTS OF NANOSCIENCE AND NANOTECHNOLOGY

I. Feller, The Pennsylvania State University

The history of predictions about the societal and economic impacts of promising new technologies is replete with predictions that incipient advances will amount to little, only to have them substantially transform daily life, and those about major advances that subsequently fizzle.

Here are some brief examples from *The Experts Speak*, compiled by Cerf and Navasky (1984):

Failing to See the Future:

"When the Paris Exhibition closes, the electric light will close with it and no more will be heard of it." (Erasmus Wilson, Oxford University, 1878)

"I think there is a world market for about five computers." (Attributed to Thomas Watson, 1943)

"There is no reason for any individual to have a computer in their home." (Ken Olson, 1977)

Seeing a Future that Wasn't:

"(A) few decades hence, energy may be free — just like the unmetered air." (John von Neuman, 1956)

"(I)t can be taken for granted that before 1980 ships, aircraft, locomotives and even automobiles will be atomically fueled." (General David Sarnoff, 1955)

These quotes are more than an academic parlor game. They point to fundamental difficulties in predicting the what, where, when, and how of asserted major scientific and technological advances, however carefully and thoughtfully crafted the projections. Reasoned agnosticism is thus a justifiable intellectual starting point.

Recognizing that inherent in forecasts of revolutions is the premise that "things will be different this time," this stance leads toward reliance on the research literature to construct an analytical framework. Economists, as well as scholars in other fields, have long studied the generation, diffusion, and impacts of scientific and technological innovation. Findings from this body of research do not themselves constitute predictions about the future of nanoscience and nanotechnology, but they do outline the variables likely to determine the rate and direction of these impacts and to identify relevant research questions.

What follows is a brief distillation of mainstream propositions about the economics of scientific discovery and technological innovation, customized to nanoscience and nanotechnology. Particular emphasis is placed on what have recently been labeled "general purpose technologies," — that is, enabling technologies that open up new opportunities (and create new production discontinuities) across a swathe of economic sectors and production relationships. Although not explicitly articulated as such, each of these propositions constitutes a researchable question about the direction (but not magnitude) of public sector support for nanoscience and the rate and direction of the technical development and commercial deployment of nanotechnology.

Uncertainty

Scientific and technological advances are characterized by both knowledge and economic uncertainties. Analysis of ways in which lacunae in scientific and technological knowledge constrain the generation of the societal impacts (both positive and negative) claimed for nanoscience and nanotechnology would appear to be best left to scientists and engineers rather than to social scientists. However, a building body of case histories, highlighted in Frances FitzGerald's account of questionable scientific claims of early (and current) proponents of the Strategic Defense Initiative and in current accounts about the feasibility of the National Ignition Facility, calls attention to systematic biases toward exuberant and at times self-serving forecasts by champions of new scientific and technical approaches of their ability to solve problems. The character of these claims warrants scrutiny by social scientists (and others, of course). How close, at what cost, and at whose expense the transformation of nanoscience is to commercially feasible nanotechnology, for example, would seem to be a set of first-order questions.

Nanoscience and Nanotechnology Linkages

Nanoscience and nanotechnology provide an excellent testbed case to study increasingly commonplace statements about the blurring of distinctions between science and technology and the speed at which new scientific findings are transformed into commercially important technological innovations. Current developments at the frontiers of research in these domains also provide a natural experiment to assess alternative models (e.g., linear, pipeline models; chain-link models; Pasteur's Quadrants, soccer games) of relationships between scientific and technological advances.

Hedonic Characteristics and Demand Elasticities

The promise of nanotechnology is its ability to do some things that cannot be done by current technologies and to do some things "better" than are provided for by existing and latent technologies. Better relates to technical performance: smaller, faster, stronger, safer, reliable, even cheaper. The projected impacts of nanotechnology writ large thus represent the summed demand for these performance enhancements across the several potential uses claimed for it. Demand curve characteristics relate to projections about the values assigned by consumers to the specific performance attributes of the new technology relative to its price and that of alternative technologies.

The projected uses of nanoscience and nanotechnology encompass public sector (e.g., space travel) and private sector (e.g., manufacturing) final demands. Differences in the characteristics of demand schedules between the two sectors may affect the direction of scientific research and innovative endeavors. Public sector demand, for example, may be expected to emphasize performance characteristics, as in defense and space research, and thus to be highly price inelastic. One might thus expect the early uses of nanotechnology to be in those products produced to meet public sector demand. (In addition, these uses may reduce technical and production uncertainties through the effects of learning by doing, and thus lower the cost schedules of firms seeking to produce goods destined for the private sector market.) How these projections of market entry in turn feed back upon the posing and priorities of scientific questions is another analytical and policy-relevant question that warrants study.

In terms of the private sector, an initial conjecture is that nanotechnology goods and services will be introduced into those markets (e.g., medicine) where performance characteristics dominate alternative techniques and for which demand is highly price inelastic. Relatedly, one would expect that the impacts of nanotechnology would be longer delayed and possibly less dominant in those markets where alternative techniques were competitive on a performance basis, and consequently, where demand was more price elastic (e.g., agriculture).

Supply-side Determinants

Nanotechnology is presented as having potentially pervasive benefits — materials and manufacturing, nanoelectronics and computing, medicine and health — to cite but a few of the areas of use from the National Technology Initiative report and Drexler's *Engines of Creation*. Pervasive relates to possibilities; not all possibilities are equally (potentially) profitable, however. The timing and sequencing of where nanotechnology's impacts are felt, in part, is a market phenomenon. Different cost considerations enter into making nanotechnology technically and commercially viable for different end users. As demonstrated in Griliches's classic study on the adoption and diffusion of hybrid corn (and Brown's work in the geography of diffusion), suppliers rationally order the sequences of markets for which they customize a general purpose technology.

Creative Destruction and Partial Obsolescence

Forecasts of the puissant impacts of new technologies often are voiced in tones of Schumpeter's metaphor of creative destruction — gale-force events in which a new technology quickly and completely replaces earlier "obsolete" techniques. In fact, the process of displacement can be both slow and incomplete, with use of earlier technologies continuing to be economically rational for extended periods rather than indicative of conservative or laggard behavior. Diffusion of the new technology, and thus the range and magnitude of its societal impacts, represents displacement. In part, displacement is propelled by continuous advances in the new technology that expand the technical range of its use and possibly lower its relative price (e.g., steam-powered vessels; the Draper loom). In part, though, the process is slowed by renewed attention to improving the range of uses of earlier technologies and/or relative price reductions in this technology that preserve for it markets in which price rather than performance is a dominant selection criterion (e.g., sailing vessels).

Learning by Using

Perhaps the greatest difficulty in predicting the societal impacts of new technologies has to do with the fact that once the technical and commercial feasibility of an innovation is demonstrated, subsequent developments may be as much in the hands of users (through what Rosenberg has termed learning by using) as in those of the innovators. Consider how the Internet has rapidly progressed from a technology supported by ARPA to facilitate communications among universities with ARPA contracts and to experiment with digital communications systems, to a means by which teenagers and college students exchange music files. In the process, the societal issues have changed from involvement of universities in defense-related research to legal suits over intellectual property rights.

Complementary Technologies and Network Effects

The diffusion and impact of technological innovations is often as much a function of the development of complementary technologies and of a network of users as it is of the introduction of the discrete technology. The impact of railroads on economic growth relates to the iterative pressures that development in each component — engines, rails, brakes, signals, organizational structure — placed on the other; the value of the telephone or Internet is a function of the number of users connected to the system as well as to the number of computers that can be accessed for information. The technical or economic specification of either complementarities or network economies for nanotechnology appears to be at an early stage.

Technological Presbyopia

To paraphrase Paul David, the shift from one technological regime to another is a journey, not an arrival. To further quote from his "Computer and Dynamo," "many intricate societal and institutional adjustments, transcending in complexity and uncertainty the redirection of private investment planing, are usually entailed in effecting the passage from one "technological regime" to another. On this view there are likely to be many difficulties and obstacles that normal market processes cannot readily overcome."

What appears to be growing consumer resistance to genetically modified foods should serve as a cautionary example that the world does not necessarily beat a door to those who build a better mousetrap, however small it is. What needs to be better understood are the reasons for acceptance, resistance and rejection, both economic and non-economic.

References

Brown, Lawrence A. 1981. *Innovation Diffusion: A New Perspective.* New York: Methuen.

Cerf, Christopher, and Victor Navasky. 1984. *The Experts Speak.* New York: Pantheon Books.

David, Paul. 1991. "Computer and Dynamo: The Modern Productivity Paradox in a Not-too-Distant Mirror," *Technology and Productivity: The Challenge for Economic Policy* (Paris: OECD).

Drexler, K. Eric. 1986. *Engines of Creation.* Garden City, N.Y.: Anchor Press/Doubleday.

FitzGerald, Frances. 2000. *Way Out There in the Blue: Reagan, Star Wars, and the End of the Cold War.* New York: Simon & Schuster.

Griliches, Zvi. 1957. "Hybrid Corn: An Exploration in the Economics of Technological Change," *Econometrica*, 25 (4), pp 501-522.

Griliches, Zvi. 1960. "Hybrid Corn and the Economics of Innovation," *Science*, 132, pp 275-280.

Mass, William. 1989. "Mechanical and Organizational Innovation: The Case of the Draper Loom," *Business History Review*, 63 (Winter).

Rosenberg, Nathan. 1994. *Exploring the Black Box: Technology, Economics, and History.* New York: Cambridge University Press.

Schumpeter, Joseph Alois. 1942. *Capitalism, Socialism, and Democracy.* New York: Harper.

Stokes, Donald E. 1997. *Pasteur's Quadrant: Basic Science and Technological Innovation.* Washington, D.C.: Brookings Institution Press.

THE STRATEGIC IMPACT OF NANOTECHNOLOGY ON THE FUTURE OF BUSINESS AND ECONOMICS

J. Canton, Institute for Global Futures

Introduction

The rapid evolution of advanced technology has constantly served up innovation after innovation in super-compressed time frames — from the mapping of the Human Genome and cloning to supercomputers and the Internet. Information technology is now responsible for as much as one-third of the U.S. Gross National Product. This is an astounding metric validating we are entering an era driven by accelerated technology developments, that have increasingly a significant economic value. The rapid advance of new technology has moved beyond our ability to accurately forecast with precision the impact on economics, business and society. We need to approach this challenge with new predictive models that are designed for the real-time complex changes that emerging technologies are influencing. This is perhaps most relevant given the challenges of nanotechnology.

We are in the midst of a large-system paradigm shift driven by accelerated exponential growth of new technology. We are witnesses to faster, more comprehensive change shaped by new technology than any civilization in history. This is but the beginning of a new wave of technologies, such as nanotechnology, that will redefine, reshape and eventually transform economies and societies on a global scale. Nanotechnology is a continuation of the next chapter in the acceleration of advanced technology and, perhaps more importantly, it may point towards the transformation of the future global economy.

Nanotechnology may become an essential large-systems strategic competency that will require coordination among all sectors of society in order to become a force for enhanced social productivity. This technology is fast emerging. Nanotechnology may well shape the sustainability and wealth of nations, organizations and entire industries in the future. A central concern here is the necessity for us, together as a nation, to plan today to meet the readiness challenges that most certainly will lie ahead.

If Nanotechnology, the manipulation of matter at the atomic level, at maturity achieves even a fraction of its promise, it will force the reassessment of global markets and economies and industries on a scale never experienced before in human history. The ubiquitous nature of nanotechnology as a fundamental design science will have applications for numerous industries: manufacturing, health care, and transportation to name a few. Since we do not know what yet is possible we can only speculate on the potential. Those societies and interests that develop the next generation tools will be first to building the nanoeconomy of the 21st century.

Nanotechnology May Drive Prosperity and Global Competitiveness

Recent developments in emerging technology and its impact on business and economics would indicate that forecasts are less than accurate in predicting the future. Few would have accurately forecast innovations such as of the Internet, wireless communications or the mapping of the Human Genome. Also, there have been numerous wild forecasts that have historically seemed more like science fiction than fact. Predictions about nanotechnology have fueled the imagination. Much of this is still imagination but the future looks promising. Nevertheless, new innovations in technology are reshaping the global economy at a dizzying speed. It would be prudent to consider the possible economic outcomes given the accelerated emergence of advanced technology. To not be prepared, to spurn readiness would be unwise given the promise of nanotechnology.

It is with this in mind that we turn to nanotechnology. Why is the potential economic impact of nanotechnology so important to consider? Nanotechnology is a fundamental design science, yet to emerge, mostly theoretical today, that may well provide us with the tools to engineer inorganic and organic matter at the atomic level. Nanotechnology, if even partially realized, over the next few decades has the potential to realign society, change business and affect economics at the structural level. New business models, design tools and manufacturing strategies may emerge at price points much reduced and highly efficient.

Nanotechnology will touch all aspects of economics: wages, employment, purchasing, pricing, capital, exchange rates, currencies, markets, supply and demand. Nanotechnology may well drive economic prosperity or at the least be an enabling factor in shaping productivity and global competitiveness. Again, we are free to speculate in the dawn of such a new science.

If developments in nanotechnology reach a critical mass in supplying radically innovative breakthroughs in automated self-assembly, as one example, most vertical industries will be influenced. Most industrial and post-industrial supply chains will be changed. What if the fabrication lines for making computers are reduced in costs by 50%? What if drug development and manufacturing costs are reduced by 70%? What if energy sources were not dependent upon fossil fuels? What then might the impact be if nanotechnology were applied to real cost reductions for essential goods and services that affect quality of life, health, habitat and transportation? There would be a dramatic impact on lifestyles, jobs, and economics. Most value chains, supportive linkages, alliances and channels of distribution will be altered. Institutions of learning, financial services and certainly manufacturing will be reshaped.

We must learn to ask the questions now about how nanotechnology may change our choices, affect our lifestyles, shape our careers, influence our communities — we must ask now and prepare so we may examine the implications that may shape the future we will live in together.

The issues that remain are to consider in what timeline what actions might be taken. How might we prepare as a society for these changes? Will there be radical dislocations or a smooth coordinated adaptation? We must plan for multiple scenarios. Radical nanotechnology innovations potentially unleashed on immature markets, fragile economies and a business community ill prepared for rapid post-industrial transformation would be problematic. We see today alterations driven by e-business and the Internet already causing deep change to industries and economies worldwide.

Imagine the emergence of a nanochip that tomorrow would deliver over 50 gigahertz of speed with the processing power of ten supercomputers for the price of a quartz watch and smaller than a key chain. What might the economic impact on the computer industry be overnight?

Imagine a super-strong and inexpensive material to be used for construction and manufacturing that would eliminate the market for steel and plastics. How might that influence the economy?

In a world being reshaped daily by innovations, the absurd today is reality tomorrow. But with the intimate inter-linkage of markets, industries and economies radical breakthrough technologies will have a widespread and far reaching impact — positive and negative. It is entirely possible that, just as computers and the Internet have become vital linchpins woven into the fundamental economic landscape of today's strong economy, nanotechnology will emerge as one of the key technologies that shapes the future economy. Many of the necessary factors are in place to drive this scenario: widespread potential cross-industry applications; fast track R&D; government investment. The risks in not preparing for and examining the economic and business impact are too large to ignore.

In an era of prosperity it is difficult to consider the lack of global leadership that might befall a nation such as the United States. How might the United Kingdom have better prepared for its 19th century challenges if it had known what was to come at the height of its global leadership in the last century? We might well ask the same questions today. Readiness is always a wise choice. Especially when it appears we may not need to be vigilant.

Nations today — ill prepared to capitalize on the Internet, the transformation of supply chains or the mobile commerce sparked by advanced telecommunications — are playing catch up and it has hampered their productivity, GDP and competitiveness.

The Nanotechnology in Business Study

In 1999 the Institute for Global Futures deployed a privately funded study to assess the general awareness and readiness of the business community regarding the economic and business impact of nanotechnology. A series of interviews with a broad range of business executives in health care, manufacturing, medicine, real estate, information technology, consumer goods, entertainment and financial services was conducted, and is still being conducted at this time. The Institute for Global Futures, a ten-year-old San Francisco organization advises the Fortune 1000 and government on the impact of leading-edge technology on markets, society, customers and the economy. The Institute covers telecommunications, robotics, computers, life sciences, the Internet, software, artificial intelligence and a host of other technologies and forecasts trends.

Preliminary Findings

Overall, the level of awareness and readiness is low, based on the survey results. Less than 2% indicated that they thought they knew what nanotechnology was. An additional 2% had heard of nanotechnology but could not explain what it meant. Of those surveyed, 80% agreed when nanotechnology was explained in basic terminology that this was an important technology that had the potential to affect them and their business; 45% expressed an interest in learning more about nanotechnology.

Though one could question at this time, when nanotechnology is still in its infancy, largely theoretical, why should anyone care and why would we even expect readiness? The issue is one of accelerated change and its impact on business and society. We are interested in readiness and awareness prior to the accelerated changes that may lie ahead, and not so far ahead as we might think. Important issues regarding research and development in nanotechnology are present today. There are real issues that bear examination today as we plan for the impact on tomorrow. Readiness is central to adaptation.

Nanotechnology Economic Scenarios: How Nations Prepare

In addition to this survey of business executives another activity has been undertaken as an integral part of this study. Given the relative and varying levels of social adaptation, we examined what might the potential scenarios be, given the contrasting readiness factors of a society. The following scenarios are briefly described as a way to generate further exploration and discussion. The value of these scenarios may be viewed as a catalyst for mapping future impact on an economy and society.

An attempt was made here to incorporate the key drivers that would shape the scenarios explored. Readiness is viewed as a precursor to these scenarios. The relative nature of socio-economic readiness, awareness and preparation will pre-determine these scenarios, and others yet to be envisioned here. This is a work in progress and will be updated, as new information becomes available. Societal readiness was defined as the awareness and ability to take action, it is viewed, as a mission-essential driver of economic and industrial adaptation. Readiness regarding education, capital, talent, coordination, and communications are all integrally part of the same platform. As nanotechnology may translate into the sustainability of nations, organizations and entire industries — readiness, the preparation and planning process, becomes vitally important to define and examine.

Scenario One: Brave New World (Timeline: 2020-2050)

Economic Environment: Nanotechnology comprehensively integrated into the economy due to high readiness, effective strategic planning and widespread investments by business, education, labor and government. Accelerated national policy and investments producing economic agility and rapid widespread large system change management. There is a widespread understanding of the numerous benefits from applications of nanotechnology, its strategic economic value for the nation, and its role in maintaining global U.S. leadership. Comprehensive social and industry-wide adoption has led to a positive impact on national productivity and an enhanced quality of life.

Key Characteristics: Robust gross national product; high productivity; global trade leadership; sustainable economic growth; global patent leadership; superior industrial competitiveness; integrated education and training

118

resources; strong investment climate; plentiful capital liquidity; high investment on R&D; low unemployment; high government and industry collaboration.

Future Outlook: Very positive. An ever-escalating predominance in key markets and industries leading to increased investments and innovations. An accelerated progressive and confident growth prognosis for the economy, and an enhanced quality of life for the nation. Global leadership and empowerment of third world and developing nations increasing. Accelerated investment in R&D and continued coordination with all sectors of society.

Scenario Two: Playing Catch-up (Timeline: 2020-2050)

Economic Environment: Nanotechnology partially integrated into the economy due to low readiness and inadequate strategic planning. Economy playing catch-up. Slow social and industry-wide nanotechnology adoption. Reactive cultural reaction to investment and organizational and industry leadership for accelerated national change management. Not a full commitment and investment in national nanotechnology policy.

Key Characteristics: Partial loss of leadership in key markets and industries; Lack of skilled talent; poor education and training; growing but still low investment in R&D; fragmented industry support; poor investment climate; liquidity insufficient; fragmented government and industry collaboration.

Outlook: Optimistic if rapid and strategic widespread large-systems change is undertaken in a concerted effort by business and government partnership. Difficult to regain ground in certain markets, but partial leadership in key markets is a success to be built on for the future.

Scenario Three: The Bumpy Road (Timeline: 2020-2050)

Economic Environment: Absence of comprehensive nanotechnology integration, adoption and readiness leading to a drastic reduction in post-industrial growth, poor performance in global competitiveness with a negative growth impact on the overall economy. Denial of the strategic value and importance. Inability to invest in the actions required to manage comprehensive large-system socio-economic change.

Key Characteristics: Loss of key markets and industries; rising unemployment; chaos in selected sectors; brain drain going offshore; lack of investment liquidity; low investment in R&D; fragmented business and government collaboration; flight capital moving offshore; educational support low.

Outlook: Moving forward into the future, it will be difficult to seize and attain market and industry leadership without a significant investment in R&D, education, training and private/government collaboration. A commanding market share in key industries and global leadership will have been sacrificed. Regaining this ground, certainly global leadership, will be a massive undertaking certain to strain capital and human resources. An acceptance of a less involved global leadership role will be the probable outcome.

Towards the Evolution of a Nanoeconomy and the Future Wealth of Nations

As the global economy continues to be transformed by new technology, a keen competition will develop for talent, intellectual property, capital and technical expertise. We see many of these factors responsible for shaping how nations today compete, interact and trade. Technical innovations will increasingly shape economies and market robustness. Technology will continue to drive global and domestic GDP. Competition will be fueled increasingly by fast breaking innovations in technology. Today this is obvious as rapid technological changes in telecommunications, life sciences, and the Internet demonstrates the emergence of entirely new economic and business realities. If the proliferation of today's technologies to form new business models is any indication of the speed and power of change in the economy, future nanotechnologies will make for an even more dramatic paradigm shift.

The evolution of a nano-economy, as contrasted with the petro-economy of today, is an intriguing idea. How might an economy not dependent on oil realign itself? More study will be need to be conducted in order to understand and map these scenarios. Fundamental nanotechnology innovations yet to come will set the timeline for this economic transformation. Or, nanotechnology may just become integrated into industries such as health care, manufacturing and energy much like artificial intelligence became an embedded component of new products.

In conclusion, the readiness of a nation to prepare for large-scale economic change is a challenging task. Nevertheless, the future wealth of nations, certainly the economic sustainability of nations, will be shaped by the preparations we make today. Coordinated large-systems strategic planning efforts may well shape our ability to adapt. Strategically important decisions will need to be made. Vastly important national security and economic issues lay yet unexamined. Huge cultural issues related to managing large-scale change will need to be better understood and plans formulated.

Nanotechnology provides a stimulating and somewhat awesome challenge to meet. If we had the knowledge in the 1960s and 1970s to prepare for the impact of computers or telecom in the 1990s, how might we have prepared the nation? Today we have real-time examples and a history of rapid accelerated economic change due to new technology to learn from, in preparing for the future.

It is too trite to state no one can know the future. The future may indeed be unpredictable. But we do know that without asking the hard questions, without speculating on the possibilities, without preparing the nation by building readiness, we may do ourselves a disservice that will be difficult to repair.

As nanotechnology moves from the theoretical to the practical, as many of us believe it shall do faster than is expected, then the possible impact on business, society and the economy will become evident over time. But we have a new opportunity today. Given the recent history of digital technology and access to better models of socio-economic analysis, we must consider growing readiness a social responsibility. We must consider readiness as part of our social policy.

We might well consider the possible futures that will result from our collective actions. We must have the courage to speculate on the possible nanotech futures we may shape as a nation. This will determine whether we have a Brave New World or a Bumpy Road.

NANO-SCIENCE AND SOCIETY: FINDING A SOCIAL BASIS FOR SCIENCE POLICY

Henry Etzkowitz, Science Policy Institute, State University of New York at Purchase

Introduction

The physical sciences are attempting to find a new basis for public funding, recognizing their relative eclipse by the biological sciences. The biological sciences are publicly funded on the basis of explicit or implicit promises to cure diseases although creation of new industries and jobs has also become a sub-theme in NIH budget negotiations in recent years. Finding a similar ground for the physical sciences, selecting areas of fundamental research for investment in expectation of solutions to specific social problems, was an implicit and sometimes-explicit theme of this workshop.

A social needs based strategy potentially raises the profile of the social sciences in formulating and implementing science policy. A social needs approach also brings engineering to the forefront and changes its relationship to basic research by reversing the direction of the linear model. Instead of going solely from the serendipitous results of basic research to utilization, with engineering playing the role of operationalizer, in an alternative model, engineering expertise plays the role of translating social needs into broad targets for basic research.

The social sciences are also brought into the picture to help formulate the social needs that are expected to be the basis for the development of new scientific disciplines as well as to identify unintended negative consequences of new technologies. The emerging strategy for NSF thus brings together the three elements (physical science, engineering and social sciences) that have often had an uneasy co-existence in the same organizational framework into a new cooperative relationship. In the following I discuss some of the implications of a social needs based model of science policy for nanoscience and the social sciences.

Alternative Science Policy Models

Two fundamental models of science policy were adumbrated during the early post-World War II era: (1) the run-off of useful results from the "meandering stream of basic research." This became known as the linear model which presumed a one-way flow from basic research to practical results, and; (2) a zigzag line running from identification of social needs and technologies designed to implement a solution. This latter model was expected to be supported by directed basic research targeted at technological goals with fundamental results a serendipitous outcome. Both models can be found in the 1945 Bush Report, *Science: The Endless Frontier* (Bush 1945), although in subsequent years the basic research modality largely crowded out the social needs dimension of science policy.

The technological outcomes of basic research assumed the forefront of attention due to the tremendous power of a single instance. Lise Meitner's findings, in the mid-1930s, that suggested fissionability of the nucleus of the atom became the exemplar of the linear model of basic scientific research leading to practical applications. The physical sciences achieved pre-eminent status and a public funding rationale during the early post-war through the success of the wartime Manhattan project. Other successes, such as radar, and the transistor in the early post-war period, based on other R&D models, also helped.

The alternative approach, "a reverse linear model" of placing social needs first, is not a new ideal. Spaceflight goals such as the Apollo Program are an instance of a social end inducing technological development work, as well as more fundamental research, both as a support structure and as an outcome of the technology project. As the agency with primary responsibility for the U.S. space mission, NASA has assumed responsibility not only for issues directly concerned with its mission, such as increasing university S&T capacities related to space, but also cognate issues such as improving science education and technology transfer (Lambright 1995).

The Crisis of the Linear Model

Wartime success and the promise of peacetime spillovers justified the creation of Federal agencies such as the National Science Foundation and the Atomic Energy Commission, (now the Department of Energy), which have been mainstays of support for much of the physical sciences. The shock of the funding cutoff for the Superconducting Supercollider is the proximate cause for re-evaluation of future public funding prospects in the physical science community.

Attempts to renew funding based upon an assumption of a "contract" between science and society to provide funding in exchange for long term future unexpected benefits have not worked well in recent years. Such open-ended appeals, like the one made by the President of the AAAS in 1995, have been met with skepticism. The original Bush model of the Endless Frontier was built upon a catalog of social needs, including housing and national security, which science was expected to address. However, with notable exceptions such as space and the abortive civilian nuclear energy initiative, the Endless Frontier model shifted from a focus on social needs to one of basic science as an end in itself, serendipitously spinning off unexpected useful technologies.

The Private Sector Version of a Social Needs Model

Social needs, or at least those that pass the tests of economic viability and fit with company strategy, are mainstays of private sector science policy. In recent years, such "policy" has increasingly been formulated at an industry as well as individual firm level. With respect to nanotechnology, the search for a viable alternative to silicon, with its limits in computer chip design, is a strong impetus. Moreover, the relative balance between public and private science policy making is shifting, even as the two are increasingly formulated jointly.

Industrial R&D spending having passed private spending several years ago augurs a change in the relationship between public and private science policy. This secular shift has arguably changed the context for public R&D funding. Heretofore, public spending provided the base from which private R&D took off. The contemporary situation is mixed. The classic instance, the invention of the transistor at AT&T Bell Laboratories, was based on corporate support for fundamental research in solid state physics, which was targeted at improvements in telephone switching systems.

Private capital is regaining the pre-eminent status as a source of support for scientific research that it held in the pre-war era in tandem with private foundations. Private support is increasingly important for the biological as well as the physical sciences. Such support is increasingly allocated as venture capital to new firms rather than through corporate research budgets of established firms. The potential of nanotechnology to induce a wave of start-ups will allow academic researchers and their business partners to tap into capital markets for R&D support, following their predecessors in biotechnology.

Toward A Public Sector Social Needs Model

Nevertheless, public funding is still extremely important, especially as the basis for academic research. It is responsible for maintaining the traditional knowledge flow to industry through publication and consultation. It also provides the base to leverage private funds as well as foster the more recent and widespread creation of intellectual property, which ensues from the Bayh-Dole technology transfer regime (Etzkowitz 2000).

How to secure this academic base was the fundamental issue of the workshop. After decades of living off World War II successes, physical science leaders recognize that a new tack has to be taken to gain public support. Technical advance by itself will not suffice to capture the public imagination. Touting the ability of nano devices to store the equivalent of the Library of Congress in an exceedingly small electronic space produces a "ho hum" reaction in Congress and among a public accustomed to amazing technological feats.

At the same time physical scientists realize that the success of the biological sciences has not been unalloyed. Significant opposition has arisen to technologies arising from biological research. It was mentioned at the meeting that some companies, heavily invested in biotechnology, attempted to inhibit discussion about genetically modified foods. The reaction against suppression of debate tends to leave the impression that there is something wrong with the technology itself, a conclusion which may or may not be warranted. A more open information and discussion process was called for.

Recognizing these deficits in the biological sciences model and the lessening effectiveness of their own model of seeking research support, the physical sciences seek to emulate the biological sciences while remedying the defects in the bio-science funding generation model.

The two-fold strategy is to find a basis in social need to justify physical science research, on the one hand, and ally with social scientists and ethicists to analyze and head off potential ill effects or misunderstandings of emerging technologies. Thus, for example, researchers at Sandia Laboratory have extrapolated an extension of their mission of national security to global security. Within that expanded purview they have identified the problem of potable drinking water as a target for nano-membranes.

Another theme is the physical sciences as the basis for future advance in the biological sciences. Just as physical science models drawn from bio-physics and chemistry became the basis of molecular biology and biotechnology in decades past, nanoscience is predicted to be the source of future fundamental discoveries in biology.

Identifying Social Needs

Most NSF supported work on the public understanding of sciences focuses on attitudes toward science and knowledge about science rather than the ends to which science could or should be put. Several modes of identifying social needs on which to base justifications for advances in S&T have been outlined. These include Foresight and Delphi techniques, Charettes in city planning, as well as public discussion models from the philosophy of science. The identification of technology goals could also process from social science research on human needs, e.g. Maslow's hierarchy of goals.

Opportunities for the Social Sciences

There is a three fold role for the social sciences in nano: (1) analyzing and contributing to improvement of the processes of scientific discoveries which increasingly involve organizational issues where the social sciences have a long-term research and knowledge base; (2) analyzing the effects of nanotechnology, whether positive or negative, expected and unintended, hypothetically and proactively and as they occur in real-time; (3) evaluation of public and private programs to promote nanoscience and nanotechnology.

Given that future economic growth and societal advance are increasingly based on S&T, understanding the social processes by which new fields of science and technology are formed becomes more than an esoteric scholarly interest. The classic work of Joseph Ben-David (1966) on discipline formation in the 19th century needs to be renewed to take account of changed conditions. More recently, the formation of new scientific and technological fields appears to occur not only through extension and bifurcation of single fields but through a process of hybridization of elements of various fields, some of which had little or no previous interaction with each other.

The process of field and discipline formation requires investigation through in-depth interviews and bibliometrics. A comparative series of studies would be most useful to produce theoretical generalizations as well as practical advice on how to benchmark and induce the formation of new fields.

The Relevance of Social Capital to Science

The role of "social capital" in creating the conditions for scientific advance warrants further exploration.

Mersenne, who through extensive letter writing knitted the scientific community of his era together, noted the importance of social ties in solving scientific problems in the 17th century. James Watson is a more recent exemplar of the utility of informal relations, finding clues to the construction of the DNA model among distant colleagues met in Cambridge bars.

Much research in nano involves collaborations among different laboratories in companies, between company and university researchers and government laboratories. It would be interesting to bring to bear sociological network techniques to track development of collaboration in tandem with in-depth interviews and participant observation, to capture the qualitative dimensions of such interactions (Etzkowitz, Kemelgor and Uzzi 2000).

Policy Implications

Programs of networked research initiatives could also be deliberately constructed to induce new forms of collaborations in target areas along the lines of the European Union Framework Programs. The Canadian Network Centers of Excellence provides another model of large-scale collaboration, primarily among university research groups. The U.S. Advanced Technology Program emphasizes collaborations among firms, large and small, but academic researchers can also be included. Considerable social science based evaluation studies have been conducted on these programs which could be tapped as a basis for organizational design of new programs.

Conclusion: The Endless Transition

Heretofore, the role of government with respect to civilian technology was indirect and elusive. There were earlier movements toward the civilian side in the debate over science policy at the close of World War II and in 1960s efforts to translate military and space R&D into civilian uses. However, these turned out to be false starts given the resurgence of military interests and concerns.

With the end of the Cold War and the rapid expansion of the civilian economy, the balance has fundamentally shifted. Civilian technology has been advancing more rapidly than military. Indeed, this shift is recognized by those in the military who now seek increased access to civilian technology to meet their needs.

The persisting debate between advocates of broad and narrow government funding is being resolved in the context of rising S&T budgets, at least in the bio-medical field, and to legitimate future funding increases in other areas. With the Advanced Technology Program now seeking to give out its awards in every state and new mentoring programs recently legislated to ensure that SBIR awards are more widely spread, it appears that Senator Kilgore may finally have won his point in the debate with Vannevar Bush. "Kilgore insisted on giving at least some research to universities on a geographical basis, which contradicted Bush's sense that elite schools received most government funding because they had the best people" (Zachary 1997, 233). Envisioning new uses for S&T goes beyond equitable distribution of funds as a basis for science policy.

The physical sciences are presently attempting to find a new basis for public funding, recognizing their relative eclipse by the biological sciences. The biological sciences are publicly funded on the basis of explicit or implicit promises to cure diseases, although creation of new industries and jobs has also become a sub-theme in NIH budget negotiations in recent years. Finding a similar ground for the physical sciences, selecting areas of fundamental research for investment in expectation of solutions to specific social problems, and revivifying basic k-12 education through the Internet, are emerging themes of U.S. science and technology policy.

References

Ben-David, Joseph and Randall Collins, 1966. "Social Factors in the Origins of a New Science: the Case of Psychology." *American Sociological Review.* Vol. 31 pp 451-465.

Bush, Vannevar. 1945. *Science, the Endless Frontier.* Washington D.C. U.S. Government Printing Office.

Etzkowitz, Henry 2001. "The Endless Transition: Finding a Social Basis for Science Policy." *Scipolicy: The Journal of Science and Health Policy.* Vol. 1. No.2, Spring 2001.

Etzkowitz, H., M. Gulbrandsen and J. Levitt. 2000. *Public Venture Capital: Government Funding Sources for Technology Entrepreneurs.* New York: Harcourt.

Etzkowitz, Henry, Carol Kemelgor and Brian Uzzi. 2000. *Athena Unbound: The Advancement of Women in Science and Technology.* Cambridge: Cambridge University Press.

Lambright, W. Henry. 1995. *Powering Apollo: James E. Webb of NASA.* Baltimore: Johns Hopkins University Press.

Zachary, G. Pascal. 1997. *Endless Frontier: Vannevar Bush, Engineer of the American Century.* New York: Free Press.

6.3 FOCUS ON SCIENCE AND EDUCATION IMPLICATIONS

IMPLICATIONS OF NANOSCIENCE FOR KNOWLEDGE AND UNDERSTANDING

G.M. Whitesides and J. Christopher Love, Harvard University

Introduction

Nanotechnology is a current "New New Thing" in science: an area that promises new understanding of nature, and use of that understanding to build technologies that will change the world. It has captured the attention of the public and of the government, and is beginning to attract the attention of corporations. Because it is new and exciting, it has also caused flurries of public concern.

There is no doubt that nanoscience is a breathtakingly interesting area, and one of the most exciting in modern physical science. It is also clear that the path from science to technology is just beginning to emerge, and it is too early to say whether the impact of nanotechnology on society will be revolutionary or insignificant. But, as a rule, where there is new science, there is new technology, and the U.S. can not afford to be other than a very active participant in this area.

So: what is nanoscience and what is it good for? Nanoscience is the study of systems with nanometer dimensions. The upper limit of size of a nanostructure is often taken as 100 nm, but truly new phenomena — quantum behavior, properties more closely resembling molecules than microscopic objects — usually occurs at much smaller scales: 1-10 nm. Highly developed nanostructures are most evident in biology, where a host of important subsystems — from ribosomes to viruses — have nm-scale dimensions. Chemists, working from the "bottom up" have made structures — molecules — at the bottom end of this scale for many years, and have developed great skill in placing atoms in structured aggregates with great precision. They have only recently begun to connect this synthetic technology to interests in microelectronics and related areas. Solid-state science (materials science, electrical engineering, solid-state physics) has historically worked from the "top down", by writing functional patterns into macroscopic sheets of semiconductor. One of the hopes for nanoscience and technology is that the combination of a number of areas — from both "top-down" and "bottom-up" fields, and from biology and computer science —

will create a new area and lead to major advances in both understanding of science and in applications of science in technology.

Scientific Opportunities

Quantum Phenomena. Nanotechnology will be an important part of any effort to exploit quantum phenomena. Quantum behavior — especially at room temperature — depends on small structures. Especially for microelectronics — where the wavelength and ballistic mean free path of electrons are crucial parameters — room temperature quantum behavior becomes important only at dimensions of a few nanometers. Developing science and technology that explores and exploits these behaviors will require developing methods for fabrication that operate on that scale.

There are now a number of demonstrations of quantum behavior in small systems, with examples ranging from "quantum corrals" (Manoharan, Lutz and Eigler 2000; Crommie et al. 1996) to fluorescent quantum dots (Snider et al. 1999) (and, of course, with many of the properties of molecules determined by quantum behavior); there is every reason to expect many more to emerge, since quantum behavior dominates small structures (Figure 6.23).

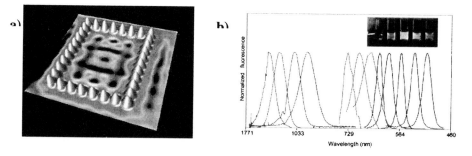

Figure 6.23. (a) Scanning tunneling micrograph of a square Fe atom corral on a Cu substrate. Reprinted from Eigler et al., © IBM (http://www.almaden.ibm.com/vis/stm/corral.html). (b) Fluorescence emission spectra of semiconductor nanoparticles of different sizes and composition (red = InAs (2.8-6.0 nm dia.), green = InP (3.0-4.6 nm), blue = CdSe (2.1-3.6 nm)). The inset shows a set of different-sized, fluorescing CdSe particles in aqueous solution excited by a single UC source. Reprinted with permission from M. Bruchez et al., Science, 281, 2013, © 1998 American Association for the Advancement of Science.

The Science of Large Numbers of Objects. Phenomena that involve very large numbers of small components follow different rules than those involving only a few components: collections of apples do not behave in the same way as collections of molecules. Nanofabrication offers the potential to synthesize small objects in very large numbers, and thus to build systems in which it is possible to examine forms of behavior that are difficult to observe when only small numbers of components interact: very large combinatorial experiments, experiments in directed evolution, and information/ computation systems based on cellular automata (Snider et al. 1999) are examples.

Atomic- and Molecular-Scale Structure and Fabrication. The combination of molecular synthesis and advanced lithography ("bottom up" and "top down" synthesis) offers a unique opportunity to fabricate nanostructures with high degrees of control. Chemistry — that is, molecular synthesis — already has enormously sophisticated tools for building highly structured collections of atoms, and chemistry routinely fabricates sophisticated nanostructures — that is, molecules — one atom at a time in quantities of tons. What chemistry has not provided is methods of building these structures in ways that are electronically or optically functional in ways that are analogous to those required in computation or telecommunications. Advanced lithography can manipulate materials — semiconductors, metals — relevant to electronics, but is cumbersome when applied to structures with dimensions below 100 nm. The opportunity to combine the sophisticated techniques of these two areas is one of the most exciting opportunities in nanofabrication.

Nanoscale Materials: Quantum Dots, Magnetic Materials, Buckytubes, Others. One of the most rapidly emerging applications of nanotechnology is the fabrication of functional structures with nanometer dimensions. Carbon-based systems (C60, nanotubes) (Rinzler et al. 1998; Franklin and Dai 2000; Dai 2000), semiconductor colloids as fluorophores for bioassays (Moronne et al. 1999; Bruchez et al. 1998), shaped semiconductor nanocrystals (Manna et al. 2000; Peng et al. 2000), FePt and Co nanoscale particles (in ordered arrays) for magnetic information storage (Sun and Murray 1999; Black et al. 2000; Murray et al. 2000; Sun et al. 2000) — all are current examples (Figure 6.24). There will almost certainly be many more to come. The techniques of small particles — nanocrystals, emulsions, colloids, micelles — are highly developed in chemistry and materials research, but have not been applied to materials that can be integrated into electronic and optical systems.

The Cell. Understanding the cell is one of the great challenges of biology, and of modern science. To understand the cell, it will be necessary to examine it using probes that are small enough that they do not perturb it (or perturb it as little as possible). Since mammalian cells are usually tens of µm in size, and bacterial cells are usually 1-5 µm in size, probes should be small on this scale. Nanostructures will almost certainly have important applications in studying cells — as intracellular probes, as tools for stimulating the cell and for measuring its responses electrically, magnetically, and optically, and in devices for manipulating the cell in sophisticated ways.

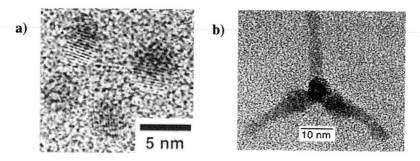

Figure 6.24. TEM images of (a) FePt colloidal particles (Reprinted with permission from H. S. Sun et al., Science, 287, 1989, © 2000 American Association for the Advancement of Science.) and (b) a CdSe tetrapod. (Reprinted with permission from L. Manna et al., J. Am. Chem. Soc., 122, 12700. © 2000 American Chemistry Society.)

Why Now? Tools and Understanding

One of the reasons that nanoscience is a rapidly expanding area is the rapid emergence of tools for study of nanoscale phenomena, and for fabrication of nanoscale structures.

Scanning Probe Microscopies. The scanning probe microscopies — of which there are now many variants — have revolutionized nanoscience, by providing the highest-resolution methods now available for determining the structures of surfaces (Stranick et al. 1996; Stipe et al. 1998).

Single-Molecule Studies. Methods for examining the behavior of single molecules are exploding (Moerner and Orrit 1999; Weiss 1999; Mehta et al. 1999; Lu et al. 1998). The most highly advanced are fluorescence-based optical methods, and scanning microscopic methods, but others — especially SEM — are also useful.

Nano-Scale Materials. Synthetic methods leading to nanostructures have been available for many years in chemistry and materials science, but the fact that these methods provided nanostructures was, in a sense, often incidental to other objectives. It is now appropriate to consider these methods as a treasure of highly developed synthetic methods capable of being applied to current objectives in nanoscience.

Atomic-Level Structure: Electron Microscopy, x-ray Diffraction, Nuclear Magnetic Resonance Spectroscopy. A variety of techniques are capable of supplying atomic-level information about structure in appropriate systems.

Molecular Synthesis. Molecular synthesis has become an enormously sophisticated field, with the ability to build an enormous diversity of molecular structures. The technology of chemical synthesis is available for use in the synthesis of nanostructures (Whitesides et al. 1991).

Meso-scale Synthesis. Extension of concepts from molecular-scale synthesis to the assembly of aggregates at larger scales is just beginning, but shows great promise. The assembly of crystalline lattices of ~100 nm-scale spherical beads is routine, although the number of structures is limited (Gates et al. 2000; Hayward et al. 2000; Ramos et al. 1999). Two-dimensional structures with high order — liquid crystals, self-assembled monolayers (Parikh et al. 1997; Allara 1995; Laibinis et al. 1989; Wilbur and Whitesides 1999), colloidal crystals (Murray et al. 2000) — are examples of ordered structures (Figure 6.25).

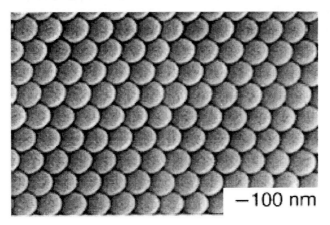

Figure 6.25. Colloidal crystal of 100 nm polystyrene spheres. (Reprinted with permission from B. Gates et al., *Adv. Mater.*, 12, 653. © 2000 Wiley-VCH, STM.)

Biology: Genomics, Proteomics, and Cell Biology. Molecular biology has been the field that has probably experienced the most revolutionary change in the last 50 years. Biology now provides an almost unlimited range of nm-scale problems, a growing number of examples of nm-scale phenomena, and a need for nm-scale probes and devices. The opportunities and problems in biology and biochemistry range from molecular issues (how to determine the structure of proteins when only a few molecules are available; how to visualize the molecular-level functioning of the cell) to larger scale problems (how to understand and exploit the function of complex organelles such as ribosomes and molecular motors).

Technological Motivation

The opportunities in nanostructures are very rich, when viewed from the vantage of science. The arguments are equally compelling from the point of view of technology.

Extensions of Moore's Law. The continuing decrease in the size and cost of transistors, and the corresponding increase in the their density, in the capability of the devices that are fabricated from them, have been major driving forces for the economy. There is general concern that current technologies are "running out", and that the rate of technological advance will be slower in the future than in the past. Microlithography using light and electrons has shown a remarkable ability to make smaller and smaller structures, and there is no reason to consider that the decrease in size of the last decades will not continue, at least for a while. Smaller sizes will, however, be more and more expensive. Moreover, as devices approach sizes of 20-50 nm, new materials issues become important. Understanding what is and is not possible in nanoscale information processing and storage devices is a key technological imperative. If it proves to be possible to go to substantially smaller devices with acceptable quality and cost, the devices fabricated using those technologies will have global impact.

Quantum Technology. A number of applications for nanoscale structures exist. Semiconductor quantum dots are being used very successfully as fluorescent labels for biological systems, and have the advantage — relative to molecular fluorophores — that they resist photobleaching (Moronne et al. 1999; Bruchez et al. 1998). Superparamagnetic particles are used for contrast enhancement in clinical applications of magnetic resonance imaging (Bonnemain 1998); different compositions have been proposed for use in information storage (Sun and Murray 1999; Black et al. 2000; Sun et al. 2000). With small size unquestionably comes new materials properties, and the potential for new applications. The search for real applications of nanotechnology continues, however, because the field has not yet identified areas in which it can contribute unique technologies.

Biomimetic or Biologically-Based Devices. Biology provides an enormous stimulus for nanoscience. The cell is based on the operation of "nanomachines" of many types — from relatively simple catalysts (enzymes) to much more complex systems — the ribosome, the light-harvesting apparatus of photosynthetic plants, the Golgi apparatus, components involved in DNA replication and in mitosis, the rotary and linear motors (Vale and Milligan 2000) that are ubiquitous in cells. The significance of these biological systems for the development of nanotechnology in the short term is not entirely clear. At worst, they provide existence proofs for molecular machines so much more sophisticated than we can presently design that simply understanding their principles of operation will keep biochemists and nanoscientists occupied for decades. At best, they may provide suggestions of designs for nanodevices, strategies for using nanostructures in new types of functions, and perhaps even components for new types of devices (Soong et al. 2000).

Databases and Information Storage. One of the early types of devices to reach the prototype stage in nanotechnology are very high-density information storage devices. The most advanced of these systems are CD-like devices in which the size of a bit is ~50 nm, and in which writing, reading, and erasing are achieved with the tip of a scanning probe device (Vettiger et al. 2000; Vettiger et al. 1999). The use of nanostructured magnetic materials also offers the possibility of very high densities of information. These sorts of systems will make it routine to store very large quantities of data, and to be able to search them efficiently.

Portable, High-Performance Systems. With very small microelectronic systems would come the ability to fabricate new classes of devices, and to revolutionize a number of functions. The movement from large computers to desk-tops, and from desk-tops to lap-tops and personal assistants, have fundamentally changed the way that data is used. If the power of the PC or the local server could be put on the wrist or worn on a belt, the world would change again. Nanotechnology — with its potential for making functional systems that are very small — has the potential to make key contributions to the goals of dramatically reducing size, and decreasing power consumption.

Sub-Wavelength Optics. A simple type of application of nanotechnology is in the area of sub-wavelength optical devices and systems: that is, devices and systems that manipulate light using structures substantially shorter than the wavelength of this light. This type of technology has the opportunity to make a range of devices — from optical filters to switches — in nanometer-scale systems having good performance and low cost.

Functional Materials: Catalysts, Materials for Energy Storage. Nanostructures are already an important class of material in many area of energy production and storage: catalysts are often based on chemically functional nanostructures; zeolites are structures with nm-scale internal dimensions, and catalysts often have nm-scale features. Bringing these classes of structures under better control, and developing the ability to design catalytic activity, has been an ambition of this field for a decade. Although progress has been spectacular, there are still major hurdles to be leapt before rational design of catalysts and other energy-related materials become a commercial reality.

Seizing The Moment: Education And Training

It is clear that there are both major scientific opportunities in nanoscience, and major requirements for nanostructures and devices from nanotechnology. What will be required to convert science into technology in this area? The first, and key issue, is the availability of scientists and engineers able to work effectively in this new discipline. This group must also have the financial support necessarily to carry out their work efficiently. It is possible but not optimal to rely on the current system of education and training to produce nanoscientists and engineers; the characteristics of the field require, optimally, a different type of training.

Nanotechnology Spans the Conventional Disciplines. Nanoscience and technology do not fit within any of the conventional scientific disciplines. Scientists and engineers developing nanoscience and technology will have to be aware of a broader and different range of subjects than those included in the usual departmental curricula. For example, a scientist interested in nanoelectronic devices should be aware of the types of functional nanostructures found in cells, and familiar with the concepts and capabilities of colloid chemistry and self-assembly. Developing the educational system to educate and train in these areas will require substantial changes in the university, with a shift from disciplinary to multidisciplinary education being an important strategy.

Collaborative Research. Since the problems in nanoscience and technology will require a broad range of talents, collaborative research will probably be more the norm than the exception in universities, and perhaps in industry. Developing styles of research that span multiple disciplines is a growing skill in universities, but one whose importance needs constant reinforcement.

Shared Facilities. Nanoscience is "intermediate" science. It does require sophisticated physical measurement, and access to specialized devices for

136

fabrication, but generally does not require very large facilities of the types represented by synchrotrons. Certain types of equipment — particularly those used for nanofabrication using e-beam writing and related techniques, and highly disciplined cleanrooms — probably do require regional facilities. Much of nanoscience is, however, done in university laboratories, and a scanning probe microscope that is local probably has more impact on the rate of progress of the field than an electron microscope located in a different state.

Types of Training. The U.S. educational system is focused on Ph.D.-level education; postdoctoral training is considered optional, and primarily for those intending to go into academic jobs. Because nanoscience and nanotechnology are interdisciplinary, the time require to learn the basics of the field may be longer than that for single disciplines, and postdoctoral training may be more a more important part of the field than it is in conventional physics, chemistry, or materials science (extensive — perhaps too extensive — postdoctoral training is already an integral part of biology).

At the other end of the scale, countries such as Switzerland have recognized the importance of well-trained people at a lower level — master technicians — in moving a new technology into commercial reality. To anticipate the need for skilled technicians, the Swiss system of training includes programs explicitly focused on this group. Specialized training of this type may be an opportunity for the U.S. as well, although there is presently no U.S. institution that has this function.

Current Status; Future Outcomes

Where will nanoscience and technology take us? What will be the outcomes, if it is successful as a field? What is the current status of the fields making up "nano"?

Nanoscience is Firmly Established; Nanotechnology is Just Emerging. There is no question that nanoscience — the exploration of phenomena at the nanometer scale, and the generation of understanding of these phenomena — is exploding. The potential for movement of that knowledge into practical application is the basis for the popular enthusiasm for this area, and the anticipation presently has reached beyond the ability of the field to deliver. For the health of the area, it is important that the scientists not promise too rapid emergence of technologies from their efforts. There are many delays between science and technology, especially in areas where manufacturing requires the production of products of very low cost and very high reliability.

Materials. Nano*materials* — functional colloids, materials such as buckytubes, quantum dots, phase-separated copolymers — are already beginning to be an important part of materials science, and the first applications — of gold and cadmium selenide nanodots as parts of clinical and biochemical research reagents for example — are already established. The future will undoubtedly see the rapid growth of the area of nanosynthesis of small (<20 nm) structures using the techniques of synthetic chemistry, and the development of applications for the resulting structures. These systems will make room-temperature quantum behavior widely available, albeit in systems that are only isolated particles or collections of particles that interact only locally. Self-assembly will play an essential role in converting nano-scale particulate materials into useful components of more complex systems: positioning the very large numbers of individual particles that would make up such systems using robotics or some other external agency is only slightly less impractical than positioning molecules in a molecular material.

Tools for Research and For Metrology. An early application for new new science is often in the development of tools that enable further investigation and measurement. These tools are already beginning to emerge rapidly from nanoscience. The family of scanning probe devices is one obvious example: these devices are now making the transition from pure research tools to tools for industrial metrology, and in that capacity will allow the measurement of characteristics of manufactured systems with a precision that was unimaginable 20 years ago. Fluorescent quantum dots are being explored actively as color labels for cell biology, and will probably move rapidly into clinical diagnostics. Buckytubes are the subject of large-scale investigation, and one of their first applications is in the fabrication of new, high-resolution probes for scanning probe devices.

Nanoelectronics. A significant part of the enthusiasm for nanotechnology is based on the possibility that it will extend Moore's law for microelectronic devices to dimensions that cannot be reached economically by existing technologies for microfabrication. The apparently inexorable progression of transistors toward smaller features has already produced conventional technology that is close to production-ready at <100 nm design rules, prototypes of transistors using extensions of this technology that have gates of 20–30 nm, and devices that can operate on the basis of the movement of a single electron. Nanotechnology will, it is widely hoped, play an important role in moving these devices toward commercial reality. The development of a large-scale commercial technology from one-of-a-kind prototypes is an expensive and long-term process, but the history of technology suggests that such laboratory demonstrations anticipate commercial reality surprisingly often: what can be made one, can often be made better in multitudes.

The issues in building nanoelectronic systems are, of course, much more complex than just building small transistors. Wiring these transistors into functioning systems is one major challenge; a second is managing the interaction between devices; a third is designing systems that will adapt to the almost inevitable failure of individual devices that will occur in systems containing numbers of devices that are presently unimaginably large.

The immediate outcome of success in making information processing and storage systems on the nanoscale would be to make levels of computing power that are presently unimaginable available for tasks that can now only be imagined: machine translation of complex spoken or written text, weather and economic prediction, data mining of enormous stores of data.

Molecules are the Ultimate Nanomachines: Using Them in Nanotechnology Will Represent a Major Step in the Development of This Field. While chemical synthesis has been developed to a breathtaking level of sophistication, and while the ability of chemists to position individual atoms in exactly defined positions in very complex structures is established technology, the combination of this type of expertise with areas of technology commonly considered as nanotechnology — especially that for computation and information storage — has just begun. The area of single-molecule electronics may be the precursor of such a combination. This area is exploring the proposition that complex molecular structures might become the basis for information systems: a molecule might, for example, become a transistor, switch, or memory cell. There are both practical and conceptual problems with this vision of a future nanotechnology, and its long-term impact on real computing systems cannot presently be evaluated. Nevertheless, it seems certain that as dimensions shrink toward the molecular, and as conventional "top down" fabrication becomes exponentially more difficult, synthetic, "bottom up" approaches will grow in importance.

Issues in Policy

Nanoscience and nanotechnology is a new area that offers access to structures, properties, and behaviors that have not previously been accessible. As with all new areas of science, with the opportunity comes challenges in policy.

Public and Private Financing. The revolution in microelectronics was financed with a combination of public and private investment. Much of the early development of molecular biology — the precursor of biotechnology — was carried out in universities with public support. Much of the early development of polymer science was carried out in industry, with private financing. The U.S. has successfully used a range of different mechanisms for financing the development of new technology, depending on the perception of its importance to national security (e.g., nuclear weapons), and public concerns (cancer), and on its relevance to commercial interests (polymer technology). An important question at this stage in the development of nanotechnology is that of the appropriate role of government in accelerating its development. Since it is still very early in the development of nanotechnology, and since U.S. industry is operating on a timeline in research and development that requires rapid return on investment, public investment — focused on science — seems to be required to move the field forward rapidly.

International Competition. Unlike microelectronics, whose development the U.S. was initially able to carry out with virtually no competition, nanotechnology has become a target of greater or lesser interest to virtually every industrialized nation. Among those regions that are active in local programs in nanotechnology are western Europe (where a number of countries have active private and public programs), Japan (where it is an obvious extension of the Japanese strength in microelectronics) and Israel (which has a national enthusiasm for high technology that has deep roots in concerns for economic security). It is useful to remember that the most important inventions of nanotechnology — scanning probe microscopy, C60, giant magnetoresistive materials — were made first or simultaneously in Europe. The U.S. will not have nanotechnology to itself.

The important question in policy is to ask what national program will best ensure the rapid and productive development of nanotechnology in the U.S.? This question has components in finance, in training, in intellectual property law, in tax and technology export policy, and in the other areas of policy that touch on any new technology.

Implications for National Security. A number of areas relevant to national security have already been identified as targets for nanotechnology. Quantum computing is one possible approach to important problems in cryptography. Global information systems for the DOD will require the movement and analysis of staggering amounts of information, and the development of new tools for information management and data mining. The vision of future war fighting relies on U.S. technological superiority —

especially information superiority — to give it a key advantage and to allow it to achieve its objectives with minimized casualties. This vision requires a dramatically improved capability to manipulate information. All of these objectives will require new information systems, and nanotechnology may be able to provide at least some components of these systems.

The Role of Small Business. Biotechnology exploded in the U.S. at least in part as a result of the active participation of the U.S. system of risk capital in its development. Is this type of investment appropriate for nanotechnology, or is it intrinsically an area (like modern microelectronics) that requires management of very large capital investments, and is dominated by large companies rather than startups? This question is not one that can presently be answered, since there are so few candidate "technologies" that have emerged from nanotechnology. The policy for public investment should, however, be guided by the method in which the technology develops: public investment intended to assist large companies is often substantially different from that best suited for high-technology startups.

Privacy. The potential of nanotechnology to make it possible to acquire and manipulate very large sets of data raises fundamental questions about privacy. When it is possible to track every citizen, and to store information that would predict patterns of future behavior, it is important to set guidelines for the use of that information early. The development of large commercial databases using current technology is proceeding rapidly; nanotechnology has the capability to accelerate this development in the future.

Public Perception: "Grey Goo". As with many new technologies, nanotechnology is the subject of public concern. Encouraging and building a sound public understanding of the potentials and limitations of nanotechnology is essential in avoiding future misunderstandings based on misperception.

Conclusion

Nanoscience and nanotechnology represent one of the most exciting areas now being actively explored. It is, in a sense, an ultimate frontier: it extends fabrication from the current, micron, scale to the scale of atoms and molecules; there is no smaller structure that can be fabricated. The next step would be into the nucleus, where different phenomena reign. Microelectronics has provided a convincing demonstration that in some fields, "smaller is better." Nanotechnology extends that idea to "smallest is best."

Nanoscience also brings the promise of new phenomena: for example, quantum behavior and biomimetic systems. In these new areas may lie opportunities for scientific and technological revolutions. With revolution comes unease by those who are observers rather than participants, and it is essential that a program of public education build an understanding of the potentialities of nanotechnology sufficient to defeat unwarranted claims of risk (as for "grey goo").

The development of nanoscience, and its conversion into nanotechnology, will be carried out with intense international competition. If there are vital commercial technologies that will emerge from this area, it is essential that the U.S. be the leader in the field. To do so will require it to establish a clear vision of what can be achieved, and to provide both the environment for research and the people able to execute the research that are required for rapid progress. It is also important to monitor the rate at which technology is picked up by the private sector, and the types of organizations that perform best, and tailor public investment to match the output of the U.S. research universities to these organizations.

References

Allara, D. L. 1995. "Critical Issues in Applications of Self-Assembled Monolayers", *Biosensors & Bioelectronics*, 10, 771-783.

Black, C.T.; Murray, C.B.; Sandstrom, R.L.; Sun, S.H. 2000 "Spin-dependent tunneling in self-assembled cobalt-nanocrystal superlattices", *Science*, 290, 1131-1134.

Bonnemain, B. 1998. "Superparamagnetic agents in magnetic resonance imaging: Physicochemical characteristics and clinical applications - A review", *Journal of Drug Targeting*, 6, 167-174.

Bruchez, M.; Moronne, M.; Gin, P.; Weiss, S.; Alivisatos, A.P. 1998. "Semiconductor nanocrystals as fluorescent biological labels", *Science*, 281, 2013-2016.

Crommie, M.F.; Lutz, C.P.; Eigler, D.M.; Heller, E.J. 1996. "Quantum interference in 2D atomic-scale structures," *Surface Science*, 362, 864-869.

Dai, H. J. 2000. "Controlling nanotube growth," *Physics World*, 13, 43-47.

Empedocles, S.; Bawendi, M. 1999. "Spectroscopy of single CdSe nanocrystallites," *Accounts of Chemical Research*, 32, 389-396.

Franklin, N.R.; Dai, H.J. 2000. "An enhanced CVD approach to extensive nanotube networks with directionality", *Advanced Materials*, 12, 890-894.

Gates, B.; Park, S.H.; Xia, Y.N. 2000. "Tuning the photonic bandgap properties of crystalline arrays of polystyrene beads by annealing at elevated temperatures", *Advanced Materials*, 12, 653-656.

Hayward, R.C.; Saville, D. A.; Aksay, I.A. 2000. "Electrophoretic assembly of colloidal crystals with optically tunable micropatterns", *Nature*, 404, 56-59.

Laibinis, P.E.; Hickman, J.J.; Wrighton, M.S.; Whitesides, G.M. 1989. "Orthogonal Self-Assembled Monolayers - Alkanethiols On Gold and Alkane Carboxylic-Acids On Alumina", *Science*, 245, 845-847.

Lu, H.P.; Xun, L.Y.; Xie, X.S. 1998. "Single-molecule enzymatic dynamics", *Science*, 282, 1877-1882.

Manna, L.; Scher, E.C.; Alivisatos, A.P. 2000. "Synthesis of Soluble and Processable Rod-,Arrow-, Teardrop-, and Tetrapod-Shaped CdSe Nanocrystals", *J. Am. Chem. Soc.*, 122, 12700-12706.

Manoharan, H.C.; Lutz, C.P.; Eigler, D.M. 2000. "Quantum mirages formed by coherent projection of electronic structure", *Nature*, 403, 512-515.

Mehta, A.D.; Rief, M.; Spudich, J.A.; Smith, D.A.; Simmons, R.M. 1999. "Single-molecule biomechanics with optical methods", *Science*, 283, 1689-1695.

Moerner, W.E.; Orrit, M. 1999. "Illuminating single molecules in condensed matter", *Science*, 283, 1670-1676.

Moronne, M.M.; Ben Dahan, M.; Bruchez, M.; Hamamoto, D.J.; Weiss, S.; Alivisatos, A.P. 1999. "Nanocrystal phosphors as biological probes for fluorescence microscopy", *Biophysical Journal*, 76, A450-A450.

Murray, C.B.; Kagan, C.R.; Bawendi, M.G. 2000. "Synthesis and characterization of monodisperse nanocrystals and close-packed nanocrystal assemblies", *Annual Review of Materials Science*, 30, 545-610.

Parikh, A.N.; Schivley, M.A.; Koo, E.; Seshadri, K.; Aurentz, D.; Mueller, K.; Allara, D.L. 1997. "n-alkylsiloxanes: From single monolayers to layered crystals. The formation of crystalline polymers from the hydrolysis of n- octadecyltrichlorosilane", *Journal of the American Chemical Society*, 119, 3135-3143.

Peng, X.G.; Manna, L.; Yang, W.D.; Wickham, J.; Scher, E.; Kadavanich, A.; Alivisatos, A.P. 2000. "Shape control of CdSe nanocrystals", *Nature*, 404, 59-61.

Ramos, L.; Lubensky, T.C.; Dan, N.; Nelson, P.; Weitz, D.A. 1999. "Surfactant-mediated two-dimensional crystallization of colloidal crystals", *Science*, 286, 2325-2328.

Rinzler, A.G.; Liu, J.; Dai, H.; Nikolaev, P.; Huffman, C.B.; Rodriguez-Macias, F.J.; Boul, P.J.; Lu, A.H.; Heymann, D.; Colbert, D.T.; Lee, R.S.; Fischer, J.E.; Rao, A.M.; Eklund, P.C.; Smalley, R.E. 1998. "Large-scale purification of single-wall carbon nanotubes: process, product, and characterization", *Applied Physics a-Materials Science & Processing*, 67, 29-37.

Snider, G.L.; Orlov, A.O.; Amlani, I.; Zuo, X.; Bernstein, G.H.; Lent, C.S.; Merz, J.L.; Porod, W. 1999. "Quantum-dot cellular automata: Review and recent experiments (invited)", *Journal of Applied Physics*, 85, 4283-4285.

Soong, R.K.; Bachand, G.D.; Neves, H.P.; Olkhovets, A.G.; Craighead, H.G.; Montemagno, C.D. 2000. "Powering an inorganic nanodevice with a biomolecular motor", *Science*, 290, 1555-1558.

Stipe, B.C.; Rezaei, M.A.; Ho, W. 1998. "Single-molecule vibrational spectroscopy and microscopy", *Science*, 280, 1732-1735.

Stranick, S.J.; Atre, S.V.; Parikh, A.N.; Wood, M.C.; Allara, D.L.; Winograd, N.; Weiss, P.S. 1996. "Nanometer-scale phase separation in mixed composition self-assembled monolayers", *Nanotechnology*, 7, 438-442.

Sun, S.H.; Murray, C.B. 1999. "Synthesis of monodisperse cobalt nanocrystals and their assembly into magnetic superlattices", *Journal of Applied Physics*, 85, 4325-4330.

Sun, S.H.; Murray, C.B.; Weller, D.; Folks, L.; Moser, A. 2000. "Monodisperse FePt nanoparticles and ferromagnetic FePt nanocrystal superlattices", *Science*, 287, 1989-1992.

Vale, R.D.; Milligan, R.A. 2000. "The way things move: Looking under the hood of molecular motor proteins", *Science*, 288, 88-95.

Vettiger, P.; Brugger, J.; Despont, M.; Drechsler, U.; Durig, U.; Haberle, W.; Lutwyche, M.; Rothuizen, H.; Stutz, R.; Widmer, R.; Binnig, G. 1999. "Ultrahigh density, high-data-rate NEMS-based AFM data storage system", *Microelectronic Engineering*, 46, 11-17.

Vettiger, P.; Despont, M.; Drechsler, U.; Durig, U.; Haberle, W.; Lutwyche, M.I.; Rothuizen, H.E.; Stutz, R.; Widmer, R.; Binnig, G.K. 2000. "The "Millipede" - More than one thousand tips for future AFM data storage", *IBM Journal of Research and Development*, 44, 323-340.

Weiss, S. 1999. "Fluorescence spectroscopy of single biomolecules", *Science*, 283, 1676-1683.

Whitesides, G.M.; Mathias, J.P.; Seto, C.T. 1991. "Molecular Self-Assembly and Nanochemistry - a Chemical Strategy For the Synthesis of Nanostructures", *Science*, 254, 1312-1319.

Wilbur, J.L.; Whitesides, G.M. 1999. *Self-Assembly and Self-Assembled Monolayers in Micro- and Nanofabrication*; Timp, G., Ed.; Springer-Verlag New York, Inc.: New York, pp 331-370.

NANOTECHNOLOGY, EDUCATION, AND THE FEAR OF NANOBOTS

R.E. Smalley, Rice University

The National Nanotechnology Initiative is a vital step toward reinvigoration of our nation's youth for careers in science and technology. Technology at the nanometer scale where we strive to build in Nature's way at the ultimate level of finesse, one atom at a time, offers our best hope of alleviating human suffering, solving the most vexing of worldwide environmental problems, and raising the standard of living of the burgeoning global population through technical innovation and economic growth. The combination of high tech gee whiz, high social impact, and economic good sense gives the dream of nanotechnology the ability to inspire our nation's youth toward science unlike any event since Sputnik.

Yet there are concerns. Some wonder that the power of nanotechnology may be so great that becomes both its own, and humanity's, undoing. Such fears are deeply embedded in our culture, reaching back to the oldest myths of the Garden of Eden and the Forbidden Fruit. Now in the millennial year 2000 the principal fear is that it may be possible to create a new life form, a self-replicating nanoscale robot, a "nanobot." Microscopic in size, yet able to be programmed to make not only another copy of itself, but virtually anything else that can be imagined, these nanobots are both enabling fantasy and dark nightmare in the popularized conception of nanotechnology. They would enable the general transformation of software into atomic reality. For fundamental reasons I am convinced these nanobots are an impossible, childish fantasy. The assembly of complex molecular structures is vastly more subtle and complex than is appreciated by the dreamers of these tiny mechanical robots.

We should not let this fuzzy-minded nightmare dream scare us away from nanotechnology. Nanobots are not real. Let's turn on the lights and talk about it. Let's educate ourselves as to how chemistry and biology really work. The NNI should go forward both here in the U.S. and in major research programs around the planet.

MATHEMATICAL CHALLENGES IN NANOSCIENCE AND NANOTECHNOLOGY: AN ESSAY ON NANOTECHNOLOGY IMPLICATIONS

M. Gregory Forest, University of North Carolina at Chapel Hill

Nanoscience and nanotechnology mark a passage to a new length scale of inquiry. The journey down to the land of individual atoms and molecules promises expansive societal impact: in fundamental science, in new technologies, in engineering design and production, in medicine and health, and in education. There is widespread anticipation of new discoveries, challenges, and understanding, yet the form and content are still blurred and mysterious. This essay addresses challenges for the mathematical culture, in the context of a rapidly evolving science and technology. I am motivated by what mathematics has to gain as well as contribute.

A consensus realization is that the greatest advances await the *interactions* between engineers, geneticists, chemists and physicists, pharmacologists, mathematicians and computer scientists. The gaps between science and technology, between education and research, between academia and the marketplace, are rapidly narrowing. There are significant demands at the interfaces between traditional disciplines and cultures. The institutions, and the cultures, that embrace interdisciplinary activities will reap the discoveries and rewards of nanoscience. By cross-cultural engagement, we can preserve disciplinary excellence and identity while removing impediments to effective interdisciplinary activities.

This science and technology revolution is unique, as our colleagues from the social sciences have underscored, in that we have the opportunity this time to anticipate implications for society and act accordingly. This means we have to prepare the groundwork not only for scientific activities, but beyond that for the general public to understand, accept, and be a part of the changes coming. In the higher education community alone, we have some significant challenges if we are to prepare future generations of U.S. citizens to participate and drive this revolution. We have to lay the groundwork for our kids to play in this new playground of discovery and innovation, arguably the penultimate toy box. What will we do in our microcosm of the mathematics culture to prepare future students to bring mathematics to bear on nanoscience and nanotechnology?

As most of this essay was written in the midst of the 2000 Summer Olympics, and I am focusing on mathematical challenges amid the excitement in nanoscience and nanotechnology, the following perspective comes to mind. If any scientist, or team of scientists, were to convene a "Dream Team" for nanoscience and nanotechnology, analogous to the team assembled for the Manhattan Project, there would be significant representation from the mathematical sciences. To quote Rita Colwell, the Director of the National Science Foundation, "mathematics is the ultimate cross-cutting discipline". The visible front line of mathematics in nanoscience is *scientific computation*, now universally accepted as a critical technology in all of science and technology. Scientific computation aids in guiding and interpreting experiments, provides predictions at the scale of individual atoms and molecules based on current quantum and atomistic theory, and can reveal collective behavior of many atoms and molecules that is only witnessed at larger length scales. As mathematics has always been the common language of science, computation has become a common tool of science and a catalyst for strong interactions between mathematics and science. I emphasize this as a social implications statement, since any nanoscience team will welcome discussions surrounding their modeling and computational efforts. Computation is the liaison between experiment and theory: a theory and a mathematical model are prerequisite to a computation, and an experiment is the ultimate validation of any theory, model, and computation.

Mathematical models are the stepping stones toward a fundamental, predictive theory. Models are a fundamental link in the scientific process, and too often we fail to emphasize in our educational system the modeling phase of science and technology. A mathematical model is based on formulation of equations and inequalities from first principles, and on the current understanding of all the complex contributions to principles such as mass, momentum and energy balances. In any realistic physical system, educated compromises (i.e., simplifications, approximations) have to be made to achieve a model that is tractable. One has to either be able to say something analytically (through exact solution or qualitative properties) or be able to compute the model equations numerically. In this sense, mathematical modeling is a complex process, an intermediate step in the ultimate goal of a fundamental theory based completely on first principles and computable both accurately and efficiently.

Especially in nanoscience and nanotechnology, modeling plays a central role if we are to be able to control the outcome at a macroscopic performance level through design at the atomic and molecular scale. There simply are far too many decades of spatial and time scales to contend with, far too many degrees of freedom, with intriguing and unexplained behavior to be uncovered at every scale of observation. Mathematical models become a necessity in this venture: indispensable for interpreting experimental data, and for guiding, explaining, hopefully optimizing, experimental observation and behavior. An effective model shortens the pathway to new products, new understanding, the analog of a smart guide who self-corrects and learns from past experience. Modeling is also not the sole province of mathematics, rather another bridge that crosses the scientific culture.

Theory is critical at every stage of development in science. A state-of-the-art model of epitaxial growth of nanometer-thin (one or two molecules thick) films requires a clear formulation of assumptions that produce the model, an evaluation of the sensitivity of the model to physical process conditions, and some confidence that the equations and computational algorithms are consistent with the experimental control conditions. Theory ultimately leads to definitive conclusions and understanding of the physical system, and more often than not new mathematics has to be created to achieve understanding. Transitions of existing theory are valuable and occur frequently, without fanfare, when models are structurally similar to known systems for which mathematical rigor has been established. But in the emergent world of nanoscience and nanotechnology, many new and different models arise that challenge current mathematical knowledge. New theory occurs on unpredictable timescales, the most powerful theories cast profound implications. Conceptual breakthroughs are needed to simultaneously: design at the atomic and molecular scale; control and optimize performance of materials and devices; and mimic nature's efficiency of assembly and mass productivity. *This promise compels the interaction of the full mathematical talent pool.* Societal implications are many and varied: benefits of active mathematical engagement, balanced by major challenges to grow the mathematical infrastructure, to weave mathematical content into all educational levels, and last but not least, to integrate the mathematical community into the nanoscience and nanotechnology revolution. These challenges are so significant that no less than a concerted national effort will do. The recently announced "Mathematical Sciences Initiative" of the National Science Foundation poses as a launch pad for such a systemic action. The mathematical community must reclaim the healthy tripartite: fundamental theory, interdisciplinary mathematics and computation, and mathematical education for a literate society.

Where has mathematics contributed thus far? The major algorithms (Cipra 2000) of the applied and computational mathematics, computer science, and statistical physics communities are central to important breakthroughs already achieved, and those soon to follow. While it is natural to focus on the products of nanoscience and nanotechnology, it is incumbent upon the scientific culture to make sure the enabling infrastructure is recognized and resourced. Mathematical science falls squarely in this category, along with various other instrumentation and visualization technologies for example. Computational algorithms in nanoscience one can highlight thus far include the following:

- Fast multipole and fast summation methods — critical to current chip design codes (Senturia, Aluru, White 1997) and Ewald sums in quantum and molecular chemistry codes (Darden et al. 1999)

- Domain decomposition methods used in film spreading simulations to connect nanoscale resolution of molecular precursor layers with continuum fluid mechanics of the macroscopic scales (Hadjiconstantinou 1999)

- Acceleration methods for molecular dynamics simulations (Voter 1997)

- Adaptive mesh refinement methods, the key to the quasi-continuum method that combines macroscale, mesoscale, atomistic and quantum mechanical models within one computational tool (Tadmor, Phillips, Ortiz 2000)

- Interface tracking methods, e.g., the level set method of Osher and Sethian, critical in etching and deposition codes for semiconductor design (Adalsteinsson, Sethian 1997) and in codes for epitaxial growth (Caflisch et al. 1999)

- Energy minimization methods coupled with nonlinear optimization methods (key elements of protein folding codes (Pierce and Giles 2000)

- Control methods (applied to thin film growth (Caflisch et al. 1999)

- Multigrid methods now being integrated into electronic structure calculations and further targeted upon multi-scale macromolecular fluids (Brandt 2000)

- Advanced electronic structure methods aiming toward larger molecules than presently possible (Lee and Head-Gordon 2000)

- Yet others noted below

Allow a short social commentary. The computational mathematics community is, at this moment, shifting attention directly onto these (and related biological) challenges, in all cases drawn into the mix by virtue of collaboration with our scientific and technological colleagues. Mathematicians are not the pioneers of these revolutions, so they have to have active relationships with those who are. New national programs at these interfaces are accelerating this cross-disciplinary shift. If we can simultaneously infuse resources and emphasis at the mathematics core, this social and intellectual diversification can succeed. Rita Colwell's timing with a Mathematical Sciences Initiative is wise.

The classes of major algorithms are a technology of their own, making it possible to carry out the numerical simulation of physically inspired models and strategies. These algorithms, not to mention those so prevalent they are taken for granted (the fast Fourier transform is a prime example), are the result of entire generations of creative mathematical science. Nanoscience and nanotechnology challenge computational science to develop faster and more accurate algorithms, and to integrate many methods into a working ensemble that can as reliably as possible accommodate all the competing physics, chemistry, length scales and timescales. To pick on one promising numerical capability: the recent quantum chemistry advances in electronic structure theory calculations (e.g., Arias and Ismail-Beigi 2000; Parr and Yang 1989) and density functional theory are now entering engineering design codes. Yet their apparent remarkable accuracy is not well understood, and compels significant research by numerical analysts and theorists before we can confidently proceed to more complex systems.

Mathematical theory and modeling are critical in virtually every experiment and technology. Data is organized in the context of some model, which is based on some theory. Experiments yield observation and documentation, whereas theory yields understanding of why things behave the way they do. Models yield predictions, most often through numerical simulation, of experimental outcomes. Successful theory and models provide an optimization or control strategy, the experimentalist's dream of a platform from which to extrapolate to the next design or application. In biology these days there are "model organisms" (Dangl 1998), which are simplified versions of the real thing; in nanoscience there are analogs such as carbon nanotubes, especially in small quantities like two or three, for which one may reasonably expect to do atomistic modeling and simulation in real time, in conjunction with experiments. By doing so, we can learn the fundamental principles of model molecular systems, and perhaps learn how to fabricate, and how self-organization emerges as systems grow in size.

The fundamental understanding of simple atomic and molecular structures is the basis for *scale-up* to more complex and realistic systems, materials, and devices. Electronic and atomistic theory and computations, together with physical experiments, are routinely becoming the basis for effective potentials in larger length scale, or so-called coarse-grained models. Technology is most often driven by macroscopic performance properties, for example strength or heat transfer or electronic properties, for which there are many time-honored (and some recent) continuum theories. Yet today these macroscopic properties are being modified, often radically, through nanoscale manipulation (e.g., nanocomposites). These explorations force us to work backward from continuum models down to molecular and atomic scale models, what one might call *scale-down*. There are hierarchies of models in every field of application, each focused at particular length scales or on particular behavioral aspects of the material or device or process. We are constantly passing up and down the ladder of length scales, and building the links between rungs of the ladder. Mathematical science is critical in the developmental process of deciding what to average over, how to accommodate the unresolved length scales and timescales and physics, e.g., through effective potentials or empirical correlations in the model. Additional intriguing problems exist for those applications of nanoscience that have no aspiration to be bigger: nano-sensors of invading viruses or toxic atoms or molecules in extremely low concentrations; planting single atoms (e.g., of a metal) in molecular chains that can totally change electronic and magnetic response, alter self-aggregation topologies, or mimic photosynthesis. Explain that mathematically!

There are challenges for statistics and probability having to do with data management and information complexity, the role of randomness and uncertainty in both measurements and models, stochastic partial differential equations are natural at the level of molecular scale modeling (Larson 1999), and Monte Carlo calculations are the bread and butter of computational biochemistry. The turbulent transport community can find a range of problems, notably the compressible transport of micelles in what appear to be revolutionary solvent processes (DeSimone 2000).

Many of the exciting discoveries and nanoscale explorations probe phenomena whose descriptions are central to applied probability, applied mathematics and the statistical physics community (who have a history of interaction). In particular, critical phenomena are pervasive. For example, many technologies and processes operate in the dynamic neighborhood of unstable transitions. The disorder-to-order phase transition of liquid crystals and many macromolecular materials is present in many polymer systems. By controlling concentration or temperature, a transition to an ordered phase occurs which, depending on the molecules that make up the system, can have purely orientational order (nematics), also acquire a handed twist along an axis (cholesterics), or acquire translational order along planes (smectics) (de Gennes 1974).

A mixture of different polymeric molecules strongly prefers to phase separation over miscibility. Repulsion of dissimilar polymer chains dominates mixing for a geometric reason: molecular chains are not free to translate in all directions due to their topology (they are not spheres), so weak repulsive forces win. This natural fact, coupled with a disperse mixture of polymeric molecules, leads to many incredible structures at a variety of length scales. So-called block copolymers (just what you think, polymer A bonded to polymer B), want to phase separate but cannot break the covalent bonds. Instead, micro-domain structures, called micelles, are possible: spheres, blocks, cylinders, worms, planar lamellae, etc. These structures are nanoscale, and transitions between micellar phases occur by changing the individual blocks, their relative lengths, concentrations, etc. (Grosberg and Khokhlov 1994). Micelles also arise in surfactant (soapy) solutions, where hydrophobic head groups are attached to hydrophilic tail groups, and transitions between phases (spherical, worm-like) may be exploited for solvent and transport processes (DeSimone 2000). These supercritical solvent processes and coating technologies exploit the properties of carbon dioxide close to the liquid-gas phase transition.

With polymeric fluids, transition phenomena interact with flows and interfacial effects to generate micron-scale patterns and defects that control mechanical properties of materials (Donald and Windle 1992; Larson 1999). Modern kinetic theories derived at the molecular scale have proven quite successful (Doi and Edwards 1986) in duplicating many observed properties of complex macromolecular fluids in simple shear and extensional flows. Because material properties of macromolecular fluids are strongly flow-dependent, significant open problems remain with characterizing the basic constitutive laws for these materials, and with characterizing the dependence of properties like viscosities and relaxation spectra. The mathematics of these coupled microstructure-flow systems is quite challenging and a very active area of research. The text of Larson 1999 and references therein provide an up-to-date status report of problems and methods. These systems reproduce the basic structure of earlier continuum theories for small molecule liquid crystals attributed to Ericksen, Frank, and Leslie, but with more complexity in the orientation field and in the variety of molecular potentials.

This is an area where mathematical modeling, the tools of applied mathematics, and computational mathematics are extremely valuable. The full kinetic theory for even simple microstructures requires massive numerical expertise and software and hardware resources. The equations themselves are changing all the time as we understand more of the physics and chemistry, and the mathematics of the derivations and structure of the governing systems. Thus it is critical to analyze and evaluate each generation or hierarchy of complex fluid models, to benchmark their predictions and determine what phenomena are captured. The interplay between these investigations, experimental observation, and theory drives the evolution of the field. This role of modern applied mathematics has attracted my interest, which has centered on mesoscale, averaged theories for complex fluids. These are tensorial, reaction-diffusion systems for macromolecular orientation coupled to the fluid equations. My research group, in the context of similar investigations in the rheology culture, has contributed to basic analyses of stability of simple flow-induced orientation patterns (Forest, Wang, and Zhou 2000a), construction of exact patterns that mimic experimental observation (Goldbeck-Wood and Windle 1999; Forest, Wang, and Zhou 2000b), and numerical simulations of the emergence of mesoscale structures (Tsuji and Rey 1998; Kupferman, Kawaguchi, and Denn 2000; our yet unpublished calculations). There are complex heterogeneous structures, steady and transient, whose properties have thus far eluded satisfactory explanation and understanding.

In new holographic processes for polymer-dispersed liquid crystals (Vaia et al. 2000), another critical phenomenon, gelation, occurs early in the process as polymerization and phase separation of disperse polymer chains and liquid crystal droplets are heading toward equilibrium distributions. Each of these processes is rather well understood in isolation; their coupling is what needs to be understood, yet the actual model equations consist of a dynamic distribution of coupled reaction-diffusion equations. The model equations for each phase have been, for decades, assumed to be of Cahn-Hilliard type; recent evidence (Weinan E 2000) suggests a new mathematical construct is needed. Striking is the complexity of this nanoscale process: the distribution of n-mers grows and extends to the far tail in finite time (the gelation critical phenomenon), creating an elastic medium in which phase separation of the multiple species occurs. This is a tremendous challenge, computationally and theoretically, and also a remarkable playground for using mathematics to understand cutting edge, technological processes for next generation displays, tunable screens and windows, and laser-hardened materials.

The carbon dioxide technology platform is a spectacular innovation of my colleague Joe DeSimone and his collaborators. In the push for the design of nanoscale structures for applications ranging from advanced integrated circuits to molecular machinery, the liquid solvents typically used in processing destroy the very structures they are helping to create. This happens because most solvents have high surface energies. So DeSimone and company have addressed this technological limitation with liquid and supercritical CO_2, at temperatures and pressures where the surface tension and viscosity are precipitously low. Those features extend the applications from device fabrication to coating technologies; surface tensions and viscosities are so low they wet virtually anything. From a phenomenological point of view, the properties of materials at the temperature and pressure of the liquid-to-gas phase transition is an exciting area for mathematics. The near-equilibrium equations of state for a liquid or gas break down, the material is highly unstable so equilibrium thermodynamics are not applicable, and molecules hover near the transition between gas and liquid phases. To achieve manageable processes, DeSimone, Ruben Carbonell, and colleagues operate just far enough from criticality. Remarkable wetting properties of liquid CO_2 are exploited for coating technologies, and supercritical CO_2 is exploited for applications such as environmentally friendly dry cleaning (Figure 6.26). Both technologies demand models to control and optimize with. The wetting applications get to nanometer scale thickness, so numerical methods in the spirit of Hadjiconstantinou (1999) and Tadmor, Phillips and Ortiz (2000) will be important. The supercritical solvent applications involve compressible transport of contaminants, and will require models for turbulent compressible transport of CO_2-surfactant micelles, which themselves experience "aggregation transitions". These aggregation transitions are exploited to trap and release contaminants. In yet other applications of micellar solutions and aggregation transitions, one uses micelles as *delivery agents* to transport trapped nanostructures (e.g., drugs). The mathematics of these molecular transport processes is saturated with intrigue.

In these examples, and more to follow, various specific nanoscience problems that compel mathematics will be noted. This raises a challenge to the mathematics community to bridge to the science, engineering, and technology cultures that are moving forward at a rapid pace. This is a complex issue that the mathematics community at large needs to address. Most of these problems cannot be formulated as a standard mathematical conjecture. It is easy to see that topology, geometry, dynamical systems, control theory, statistics, probability, partial differential equations, etc. are central to these investigations. But they are fundamental and critical in the context of the science. For mathematicians to participate, major effort is needed to interface with the materials, chemistry, physics, engineering, computer science, and biological communities. Success relies upon sufficient resources, to be sure, but also upon educational strategies for students and faculty alike, increased collaborations with scientists and engineers, equality in emphasis and status for interdisciplinary mathematics, and leadership to shepherd the changes.

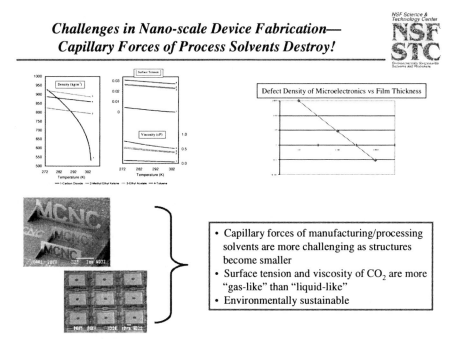

Figure 6.26. Applications of liquid and supercritical CO_2.

These pursuits require change in what scientists, engineers, health professionals, and citizens have to know simply to communicate amid this landscape. That realization will impact the strategy for our educational systems as well as reform the research enterprise among universities, government, and industry. Many of these changes are taking place, and those successes need to be advertised more and visible as models for the mathematical and scientific culture.

Two shining examples of such innovation exist at my home institution. One is the NSF Science and Technology Center on Environmentally Responsible

Solvents (website http://www.nsfstc.unc.edu/), led by Joseph DeSimone of the University of North Carolina at Chapel Hill and Ruben Carbonell of North Carolina State University, with partnerships from several other universities, private foundations, and industry. This Center integrates: nanodesign of new solvent technologies that are friendly to the environment; undergraduate, graduate and postdoctoral educational and research opportunities; and alliances among social scientists, ecologists, chemists, chemical engineers, materials physicists, computer scientists, mathematical modelers, law students, and business students. Center activities span the study of unexplained behavior of materials near critical transitions, critical technologies from dry cleaning to teflon production to new wetting and coating techniques, with the interaction of students and faculty from science, engineering, law, and business. I emphasize the prime opportunity for mathematics: design and control are limited without mathematical models and computation. Theory is far behind observation and application; what are we waiting for!

Another example is the nanoManipulator Project at UNC, a new technology that combines visualization by computer scientists (Fred Brooks and Russ Taylor), manipulation via atomic force microscopy (Rich Superfine and Sean Washburn), and materials and life science questions fed by colleagues across chemistry, physics, biology and gene therapy (www.cs.unc.edu/Research/nano). From carbon nanotubes (Colbert and Smalley 1999) to viruses, the nanoManipulator team exposes how nanoscale things respond to manipulation (Figure 6.27). They even empower middle and high school students with remote capability from their classroom to perform the same experiments the scientists can; with fast Internet II links what difference does it make where your workstation is? The students can't help but ask why those viruses or nanotubes respond the way they do; that's a question for mathematics and science. Then they are hooked with the desire to know why: that's impact! that we need to make contagious. Colleagues in education and outreach provide the ongoing linkages between the research and school enterprises.

Figure 6.27. NanoManipulator interface.

These represent two nanoscience and nanotechnology foci of intense excitement and prospect — founded in fundamental science, driving advances in technology, spawning new industrial processes, and grabbing the imaginations of the next generation. Both centers link with one another, sharing each other's innovation and instrumentation, and link to the applied and computational mathematics culture. These scientists invite collaboration on theory, modeling, computation, and algorithms, and recognize the value of and need for mathematical collaborations. It is my experience that this is the case in every academic, industrial, and government setting, in which case the upside potential for mathematics is huge. Mathematicians are participating in this scientific revolution, yet the need is orders of magnitude greater. I would challenge every mathematics department in every university to enhance existing collaborations, to seek out new ones, and to encourage faculty and students to get involved, to be part of teams of interdisciplinary focus groups in nanoscience and related life science initiatives. Department chairs should make it a priority to have liaisons from mathematics to every campus working group in interdisciplinary science. I strongly believe this is one way we can have an educational atmosphere in which mathematically talented and inclined students can pursue mathematics at varying degrees of commitment.

Another model for the interaction between mathematics and nanoscience and nanotechnology is a coherent and focused interdisciplinary and fundamental mathematics "institute" with strong anchoring to science, engineering and technology. For two decades we have had the Institute for Mathematics and Its Applications at the University of Minnesota, supported by the National Science Foundation. Similarly, the Mathematical Sciences Research Institute in Berkeley, CA is an NSF center for mathematics, providing (especially in my experience the past few years) a strong blend between pure and applied mathematics. The NSF Science and Technology Center in Discrete Mathematics and Theoretical Computer Science (DIMACS) has provided another extremely valuable venue for generating mathematical interactions with science. (I am particularly fond of the special "year" in molecular biology that started in 1994 but has yet to end.) For almost half a century, the Courant Institute has influenced applied mathematics at the interface between science and mathematical theory. The computational mathematics community owes a huge debt to Alexandre Chorin and his colleagues at UC-Berkeley and Bay-area labs for tremendous impact in populating the nation and beyond with outstanding talent; recent years have seen a new spawning ground of computational mathematicians at UCLA. The National Labs have been tremendous facilitators of interdisciplinary science, and have done (in my judgment) more to influence the interplay between mathematics and science than any other government program. My colleague Thomas J. Meyer, a National Academy chemist from UNC-CH now heading up strategic research operations at Los Alamos National Laboratory, said to me recently after less than a year at the lab: "That place is full of applied mathematicians!" He is correct, and the good news is that Meyer is happy to have the capabilities of the mathematical sciences at his disposal.

A new NSF institute resides at UCLA, the Institute for Pure and Applied Mathematics. Several other valuable institutes could have been launched were the resources available. All of the NSF supported institutes rotate emphasis around areas of mathematics, or applications that involve significant mathematics. Their mission is sufficiently broad that they cannot maintain a singular focus in nanoscience, or genomics, or any singular area of critical need and opportunity. A new competition for additional NSF mathematical sciences institutes has been announced recently, so more centers of mathematical science will emerge soon to a neighborhood near you. These centers/institutes provide beacons from mathematics to the science and technology cultures. The presentations given at the November, 2000 Mathematics and Statistics Chairs Colloquium in the nation's capital were impressive to the most cynical critic of national centers. These institutes provide a venue and catalyze activities that simply cannot and would not occur otherwise.

Yet other resource streams are necessary to encourage and sustain a dedicated infusion of mathematics in a critical scientific or health or policy challenge. At UCLA, there is a new project called Virtual Integrated Prototyping for Epitaxial Growth (website: www.math.ucla.edu/~thinfilm/). This is an interdisciplinary project on epitaxial growth, involving modeling, simulation, control, growth and sensors. The project is centered at HRL Laboratories and the Applied Mathematics Group at UCLA, with support from DARPA and the NSF, including a grant in the new Focused Research Group initiative. This is a new (and welcome) vehicle for short-term support on a targeted area of science, in this case epitaxial growth of thin films. This project is an example of substantive crossflow between mathematics and nanoscience and nanotechnology, with outcomes neither could achieve independently, that needs to be duplicated many times over. The limitations of existing algorithms and theory are exposed and addressed, from all sides of the collaboration. Such relationships are critical if mathematical issues like accuracy of ab initio methods (e.g., density functional theory) are to be acknowledged and deemed worthy of serious effort. Without such working trust, engineers and scientists will make new materials with the tools they have available and that they create, whereas our abilities and contributions might not be initiated much less brought to bear. My only complaint, albeit a big one, with the FRG initiative is that there were insufficient funds to support more than a handful of proposals. I personally know of ten or more compelling proposals from colleagues around the country that were meritorious but could not be funded. Again, the announcement of Rita Colwell of the NSF Mathematical Sciences Initiative presents the opportunity to resource many more interdisciplinary collaborations. There has to be a critical mass of such activity to affect the cultural impact necessary for mathematics to be a major player in nanoscience and nanotechnology. The mathematics community bears a responsibility if the resources indeed materialize.

I now move away from the broad discussion of applications of relatively recent mathematics in nanoscience and nanotechnology to more general discussions of what, why, and who from mathematics.

Playing with atoms and molecules became thinkable over the past century. Today, we routinely image, manipulate, design, assemble, flow, compute, and theorize — atom by atom, molecule by molecule. Nanoscience and nanotechnology combine the exploration of *matter* (materials, devices, organisms), *function* (the rules that govern how atoms and molecules behave alone and collectively), and *process* (design and fabrication, synthesis, production). Yet how do we anticipate, predict, and optimize outcomes on the atomic and molecular playground? The answer is mathematical:

159

modeling, theory, and computation. The mathematical sciences, bridging applied mathematics, pure mathematics, statistics, and computer science, cycle with observation and experiment in the modern scientific method.

One might ask, why "nano" now? The advances of the silicon age, the software and algorithm age, the connectivity instilled by the Internet, have unleashed our imagination to dare model and compute details of nanoscale matter, function, and process, perhaps to simulate in real time interactively with experiments. The physical experimentalists, engineers, health and medical professionals are forging ahead, aspiring to quantum computers, atomically designed devices and materials, and gene-based health. The spirit is palpable across the scientific and technological landscape. The passage to the nanoscale is heavily traveled.

How will mathematicians participate? Mathematics has always played a fundamental role in science: as the common language to assert models and principles; as the conceptual basis for remarkable structure in nature such as symmetries, or broken symmetries; and as the instrument for explanations, understanding, and predictions. Historically, entire fields of mathematics have been spawned in the search for understanding nature and in response to fundamental societal crises or needs. Modern mathematics, especially in the U.S., has a dual personality that can be a tremendous strength if nurtured effectively.

On the one hand, applied mathematics comprises the formulation of models from scientific principles, the development of solution methods and analyses to extract behavior of physical (and social) sciences, and the creation of numerical algorithms and their implementation for computer simulations of models. Statistics and probability play a critical role, from the design of experiments, to data analysis, to the role of uncertainty and randomness, to quantum Monte Carlo methods and kinetic theory approaches to macromolecules. Applied mathematics and statistics are the interface between pure mathematics and science, bridging theory to experiment. There is often little to distinguish the research of a computational chemist, physicist, or biologist from today's applied mathematician. The applied mathematician is more likely to worry about order of accuracy, convergence, stability, effects of under resolution, or even embark on completely new algorithms having determined serious limitations or unacceptable aspects of existing tools. The applied mathematician is more likely to explore idealized models based on isolated competitions, rather than put all the ingredients into a huge model and simulation. Pausing to formulate a hierarchy of complexity with transitions between levels is simply an instinct, sacrificing rapid rewards for balance of understanding. Thus we find ourselves naturally slowing the process of discovery in order to achieve accuracy, principles to stand on. Theoretical advances have yet another, more unpredictable, timescale. And so we participate in a climb forward, but not without some bumps in the ride. The role of mathematics is distinct, yet most often not the source of first observation or discovery, settling happily to be part of the understanding. I somehow feel that our scientific colleagues are more able to take chances and speculate, knowing that their mathematical colleagues are following close behind. Implicit in this picture is an active relationship between the applied mathematical and scientific/engineering cultures. There is a danger that our discipline will be adopted and hybridized by those who recognize its importance and need, but simply do not have a relationship with their mathematics colleagues. This is a clarion call for the mathematics culture to rise to the challenge. We simply have to replenish the interface between mathematics and science. This is not a call to compromise fundamental mathematics, rather to embrace a fuller spectrum.

On the other end of the spectrum, abstract fields of mathematics are inspirations of the human intellect, originating at some point from science or observation, or curiosity. Once the mathematical descriptions of nature or the imagination are put into place, they take on a life of their own. The fields of what we call pure mathematics are a benefit, some would argue a responsibility, of advanced society that encourages pure intellectual pursuit, unfettered by constraints of relevance. Nonetheless, many abstract mathematical creations, in the hands of those who address scientific questions, emerge to enable scientific breakthroughs. Fourier analysis (Dym and McKean 1972) is one such example, with roots in the work of Fourier in the early 1800s on the theory of heat (Fourier 1878), earlier work of d'Alembert in the mid-1700s on the theory of the violin string, and other great names from the mid-1700s to the mid-1800s: Euler, Bernoulli, Lagrange, Dirichlet, and Riemann. Scientist was synonymous with mathematician for centuries. The early 20th century witnessed the distinctly mathematical achievements of Lebesgue and Plancherel. They established the rigorous equivalence of a function and its fundamental "Fourier" modes, affirming the intuition of the early giants, but also spawning many other fields of mathematics in the process. Today we are beneficiaries of the technology of the fast Fourier transform, wavelets, and image and data compression.

The field of complex variables, in which we imagine numbers that are not real, began in the 1500s with a simple quest to extend our number system so that the general solution of cubic equations could be accomplished. It is interesting that simply writing down the square root of a negative number was not convincing; only when Bombelli reconciled the confusion over the

161

so-called irreducible case of cubics with the aid of square roots of negative numbers did the concept become plausible. This development is considered the crowning achievement of mathematics during the Renaissance, a sad comparison to other cultural contributions one might add. Nonetheless, one then naturally asks the same question for other polynomial equations, to which Euler and Gauss responded *three centuries later* with the remarkable Fundamental Theorem of Algebra: complex numbers with a real and an imaginary part suffice. But then what are the properties of *functions* of these new number systems we imagined? Onto the scene comes the work of Hardy, Littlewood, Wiener, the merger with group-theoretical concepts and representation theory, amazing new "trace formulas" that generalize the first result of Plancherel. Here the work of Frobenius, Selberg, Weil, and Weyl, and Wiles span a century of beautiful mathematics, quantum mechanics and number theory. The applications of functions of a single complex variable have been profound, and one cannot read a textbook on theoretical physics and chemistry that does not exploit the properties of complex numbers and functions. There is a mystical continuity in mathematics; the inquiries in complex and Fourier analysis touch the experiences of Andrew Wiles and his life's fascination with Fermat's last theorem. Powerful areas of mathematics were summoned by the desire to answer the next logical question after a school child learns the Pythagorean Theorem. What new angle in mathematics will emerge from nanoscience?

My own career started amid the wave of new science and mathematics surrounding solitons. Humble beginnings arose in the attempts of Korteweg and deVries (KdV) to understand long surface waves in shallow water, and of Russell to understand the solitary wave that he chased on horseback along the navigation canals in Scotland. But the breakthrough awaited development of mathematics sufficient to model and explain the Fermi-Pasta-Ulam numerical experiments in Los Alamos (Fermi, Pasta, and Ulam 1974). Ironically, the FPU study was aimed at quantifying the finite rate of heat conduction in contrast to the unacceptable result of Fourier's linear heat equation that heat is felt instantaneously at arbitrary distances. FPU surmised nonlinear lattice calculations would show a finite transfer of energy toward equipartition of energy. Instead, they observed recurrence in the lattice modes; startled, the original study was abandoned. Kruskal and Zabusky (1965) then took the ingenious step of passing to a continuum limit of the FPU lattice, and derived the KdV equation. Thirty years of soliton theory ensued, and the field of integrable systems bridged direct and inverse scattering theory, algebraic geometry, solid state physics, finite-dimensional lattices and partial differential equations, internal waves in the Andaman Sea large enough to destroy oil platforms, and most recently orthogonal

162

polynomials and optical fiber communications systems. Many physical scientists, mathematicians, and technologists have come to know one another in the aftermath of the FPU and Kruskal-Zabusky discoveries.

Where does modern mathematics enter nanoscience and nanotechnology? Applied and computational mathematicians have to be central players in these developmental years, at a significantly higher level than is currently apparent. I would recommend nothing short of a massive campaign at the departmental level on every campus to get involved with the nanoscience and nanotechnology activities. Their participation will affect the path of scientific and technological advances, as would their absence. It is critical for the health of science and mathematics that there is full engagement. The contributions of abstract mathematics are likely to be longer timescale and less predictable, but possibly profound and revolutionary. In any case, it is clear that fundamental new concepts are likely to emerge along the journey, spawning areas of mathematics that we cannot foresee. These speculations aside, a concrete look at where things are now is in order.

Physical experiments with atoms and molecules are advancing, but remain difficult and expensive. Furthermore, we remain limited in our ability to image and visualize nanoscale experiments, especially dynamic processes, and to collect data directly without disturbing the experiment. Humans of course cannot *see* nanometer structures (the human eye resolves wavelengths down to 400 nm) so we have various microscopy and scattering methods to "observe" individual atoms or get information about electronic properties. But it is still necessary to *infer* what is happening to individual atoms and molecules in most scientific and technological applications. The tool of inference, aside from divine inspiration, is mathematical modeling and computation. Models are necessarily preceded by theory that formulates the relevant physical mechanisms at play, and model predictions or computations must accompany physical experiments. The whole area of visualization in science is a critical enabling technology. Humans simply process at a far deeper level with visual input, yet visualization is not part of our training until now. Clearly converting experimental data, or data from simulations of complex models, to images is a major area of computer science. Yet it is deeply mathematical, and our discipline has to keep attention focused in these areas as well. Consider for example the beautiful geometry associated with self-assembling membrane structures (Lipowsky 1991; Michalet and Bensimon 1988) and with minimal surfaces in block copolymers (Thomas et al. 1988).

Today, modeling and numerical solution technology is highly advanced at the atomic and molecular scale. Many advances are being realized based on

163

quantum mechanical calculations (e.g., electronic structure theory, density functional theory, *ab initio* quantum mechanical simulations), yet we remain far from the scale-up to realistic numbers of atoms. This is a goal of major grand challenge initiatives in supercomputing, but we still have no choice but to construct models at larger length scales, and with fewer disparate timescales. The self-consistent removal of small length and time scales is a grand challenge to mathematics and science.

A fundamental line of questioning centers around the principles that explain *collective behavior*: of many atoms, or complex molecules, or collections of many molecules. What are the *theories* that emerge for the behavior of several atoms, complex molecules, scaling up to macroscopic properties of nano-engineered materials, devices, assembly methods? The answers will determine properties at scales from nanometers to microns to meters, which of course depend on details of the chemistry and physics of the constituent atoms and molecules, their concentrations, who they interact with, the ambient conditions. The challenges are many; the questions dominate the answers.

The only obvious conclusion one can draw at this time is that applied and computational mathematics must be a player in these developmental stages of nanoscience and nanotechnology; the mathematical community has to rise to the challenge, or else scientists and engineers will proceed without us. This does not mean that they will do what we could do; it will be a different outcome entirely, much to the disadvantage of all. Full engagement by the entire mathematical community will catalyze ideas, resources, and opportunity for mathematics within the community of science and technology. More profound is the recognition that mathematics can significantly impact the history about to unfold.

Nanoscience and nanotechnology are frontiers for every participating person, field, and institution. We bring today's understanding and tools, knowing full well that they were created for other problems and challenges. The phase we see now consists of adaptation and applications of existing technology and theory, and a beginning to identify our limitations and needs for new ideas, new tools. The imagination across science and technology is running wild. It is from this perspective that I want to offer a relatively recent analogy of the role between mathematics and science on a large-scale, societal challenge.

Historically, pressing scientific and societal crises have been met, and overcome, through collaborations between scientists and mathematicians. The Manhattan Project is one such example, and the mathematical landscape for decades after the Second World War was dominated by the fundamental limiting factors in understanding the constituent physical processes in theory, design, engineering, and detonation of atomic weapons. Functional analysis aimed at quantum mechanical issues and the theory of existence and uniqueness (or lack thereof) of the model partial differential equations, the theory of weak solutions such as shocks and rarefaction waves and selection principles from entropy conditions, the first analog computers (MANIAC) to perform bomb calculations, the first numerical algorithms to solve the compressible gas models, the theory of lattice models to mimic continuum theories and the complex relationships between discrete and continuous models, the role of turbulence in both the compressible gas dynamics and in the transport of the nuclear blast cloud, are but several of the many fields of mathematics that emanated from the Manhattan Project. I recommend the essay by Peter Lax (1989) for a perspective from one of the key players. These areas of mathematics then spawned new fields and concepts, as a natural human invention to understand the fundamental questions that had been identified. The discoveries of the past half-century then spilled over into numerous applied fields, including aerodynamics and aircraft design, environmental (marine, atmospheric, and geophysical) sciences, all areas of fluid dynamics with applications to engineering design, and in explanations of the catastrophic events occurring in the universe. Independently, areas of pure mathematics were created as second and third generation offspring of the singular commitment to the Manhattan Project.

I cannot help reflecting on individuals who, over time, have had incredible impact. When I think of a mathematician's mathematician in recent times who changed science and technology as well as mathematics, two names come to mind. Of course this is my limited view, but one of them (Peter Lax) had the torch passed to him by the other (John von Neumann) in the fallout of the Manhattan Project. Bibliographies of the genius von Neumann have been written, and I suppose one will soon be written of Peter Lax. All Peter accomplished was the modern framework for numerical solution of partial differential equations of hyperbolic type, fundamental concepts such as convergence and stability in numerical algorithms, seminal schemes such as the Lax-Wendroff scheme, all emerging from his involvement with von Neumann I suppose. Lax and Phillips developed and wrote the definitive book on the mathematics of scattering theory. But in soliton theory, it was Lax who conceived of the fundamental structure of integrable systems: the Lax pair, and the stroke of genius that a nonlinear system could be cast as the compatibility condition for two linear operators. Lax's name is attached to the entropy condition that selects the unique physical speed of shocks from an infinity of solutions. Lax and Levermore established, with full rigor,

165

the distinction between dispersive and dissipative regularizations of shocks. Imagine the impact as the next Peter Lax pays attention to nanoscience.

The current explosion in design and exploration of atomic and molecular systems (pure and composite materials, electronic and logic devices, nanoscale probes and sensors, gene-based drug design) is plenty challenging and full of mathematical openings. On the horizon there is always the mother of all challenges: to translate nanoscale design/control/understanding to performance and function at larger scales from microns to meters. This challenge is the same whether we aspire to quantum computers, gene therapy, new materials, or environmentally friendly industrial processes. A compelling recent survey (Baschnagel et al. 2000) of the state-of-the-art in modeling of polymers and macromolecular fluids is highly recommended. A related multi-disciplinary and multi-cultural approach to soft matter is underway in Japan, with a multi-year project among industry, government, and academia. The goal is development of theory, models, and numerical codes across a spectrum of lengthscales — starting at the nanoscale. This enterprise is called the Doi Project, after M. Doi, the project leader and a fundamental contributor to the microscopic kinetic theory and mesoscopic averaged theory for macromolecular fluids. Should the U.S. have analogous initiatives in nanoscience and nanotechnology? Could all the NSF Insititutes and STCs collaborate in a national program, perhaps in conjunction with national labs?

We have certain realities that must be recognized and respected. All properties of synthetic and biological systems may ultimately follow from quantum mechanics and electronic structure. Yet it is not feasible to expect computers in our lifetime to be able to calculate the complex atomic, molecular, and multi-species components that make up macroscopic objects (materials, electronic components, things biological). Even if we could perform the simulations, the data sets would swamp our ability to mine the data for information and understanding. We have mentioned above a variety of numerical successes in coarse-graining, yet clearly we need further conceptual breakthroughs, in mathematics, statistics, modeling, and computation. The mathematical disciplines will themselves undergo evolution as a result of the creations and discoveries in nanoscience and nanotechnology.

I close this essay with a selected list of additional interesting problems where mathematical and computational insights and methods provide significant pathways for involvement.

- New materials:

  - fabrication (nucleation and growth) of nanostructured materials, e.g., carbon nanotubes, issues relate to non-equilibrium thermodynamics of crystal growth.

  - nanocomposites — the mixing of nanoscale particles, of various chemical and geometric types, with traditional materials such as polymers, metals, or ceramics: what are the effective properties of the mixture with respect to strength, electronic properties, response to electric and magnetic fields, flow properties, heat transfer? These are difficult modeling, numerical, and analytical problems.

  - molecular self-assembly: the "natural" approach centered around steering nature to self assemble preferred structures, which would ameliorate the bottleneck in fabrication of sufficient quantities of many atomistic designed materials (Nauta and Miller 1999).

  - biomaterials, a perfect example where emerging ideas of self-assembly will be critical to avoid rejection by the host.

  - self-assembly is important everywhere in nanoscience, from membrane structures (Lipowsky 1991) to block copolymers (Thomas et al. 1988) to the smallest water molecules yet achieved (Nauta and Miller 1999) to micellar surfactant solutions (DeSimone et al. 2000).

  - elastomers (Warner et al. 2000; Warner and Terentjev 1996) and shape memory materials, which undergo macroscopic, reversible, strain deformations with small control variations in temperature or in response to a mechanical applied field; such materials are targeted for a variety of applications including replacement for muscle tissue.

  - epitaxial growth and thin film deposition techniques are the focus of an interdisciplinary project involving HRL Laboratories and Applied Mathematics group at UCLA. This project is an example of the kind of synergy between mathematics, science, and technology that is desirable across nanoscience. There is an emphasis on morphology at the nanometer length scale, approached through a hierarchy of models that includes atomistic, continuum and process models validated by experiments. Refer to www.math.ucla.edu/$~$thinfilm/ for further details.

- Fluidic and non-fluidic materials:

  - macromolecular fluids, such as liquid crystals, liquid crystal polymers, copolymers, where the molecular geometry and molecular weight influence intrinsic and flow-dependent behavior; incredibly

167

rich phase diagrams, with phenomena more complex than classical phase transitions (de Gennes 1974; Larson 1999).

- disparate length and time scales during formation and synthesis of macromolecular fluids distinguish properties and behavior at various length scales: molecular architecture (such as rodlike or disclike, side branches, copolymers, and so on) are realized at higher length scales as defects and patterns, some steady while others are transient; how does "theory" link these hierarchies of scales and scale-specific structures and phenomena?

- nanoscale design of "functional materials" attempts to exploit the physical (hydrogen bonds), chemical (electronic attractive and repulsive), and volumetric "crowding" forces; the length and time scales of the effective extended forces, and the functional form of these effects, are a work in progress. The action is centered on models for theory and simulation to guide experiment and design.

- the dynamics of chemistry, organization, transition behavior, structure formation, geometrically forced defects and patterns because of molecular architecture, are all applied mathematical phenomena.

- statistical and probabilistic methods and phenomena abound: phase transitions, materials processing at critical temperature and pressure, networks/gelation, all of which lead to exploration of the unknown, highly dynamic, non-equilibrium phenomena at the interface between phases of macromolecules; theory about these regimes of materials is open; simulations on any industrial scale are not possible, or are they?

- use of non-toxic solvents such as CO_2 near critical temperature and pressure for applications to wetting of surfaces, self-assembly of macromolecular structures (DeSimone et al. 2000), to mention two.

- bridging the enormous gap between atomic vs. mesoscopic vs. macroscopic time scales and length scales in pure macromolecular systems, then in composites: the need for sophisticated averaging/coarse graining methods. The mathematicians speak of averaging and homogenization, whereas the scientists speak of coarse graining. Whatever the terminology, this is the holy grail. The review article by Baschnagel et al. (2000) and the applied mathematics interdisciplinary project at UCLA on thin films (www.math.ucla.edu/~thinfilm/) are excellent sources.

- identifying the disparity between collective (or averaged) and individual nanoparticle behavior: related to problems of moments of distributions, open for new methods and ideas from statistics and probability. Experiments of Steve Chu and colleagues aim at the experience of an individual molecule in flow, and then transition behavior from dilute to concentrated regimes (Smith, Babcock, and Chu 1999).

–   numerical issues abound: disparate length and time scales force issues of resolution in numerical codes; how does one perform the analog of subgrid closure in numerical algorithms; huge efforts aimed at numerical coarse graining in polymeric systems (Baschnagel et al. 2000); polydisperse liquid crystal display manufacturing: interaction between polymerization, phase separation, and gelation lead to a broad-banded distribution of coupled reaction-diffusion equations.

–   derivation of kinetic-based theories of flowing macromolecular systems, averaging methods to get mesoscale models, and the scale-up to macroscopic bulk properties is completely wide open.

–   crystal-like structure, such as nanotubes, where the current focus is on individual or small numbers of tubes — many problems loom such as fabrication on an industrial scale, collective behavior of nanotubes alone or in mixtures; these materials are a test case for theory.

–   the understanding of the information coded in DNA, which links nanotechnology and nanoscience with the genomics, proteomics, and bioinformatics revolution. There are major challenges for statistical correlations with functional aspects of sequences. Will the translation of this information into language and grammar require new mathematical theories? Can one understand how sequence information implies function? What is the computational complexity associated with these challenges?

•   Devices are another wide open area of nanoscience and nanotechnology. Device sizes are decreasing to the point where issues that were deemed negligible just a few years ago are now fundamental, including effects of thermal fluctuations. I will not go into any detail on this topic, and only mention examples of potentially fruitful collaborations. Magnetic storage devices are a classic example where fundamental mathematical and computational challenges remain unanswered. Current tools of lattice sums and fast multipole methods are applicable, as well as analytical advances in multiscale analysis of nanostructures. This area of research involves analytical and computational mathematicians (Bob Kohn, Leslie Greengard, and Weinan E), a physicist (Andrew Kent), and industrial scientists from IBM (G. Grinstein and R. Koch). Other investigations focus on microscopic sensing devices and molecular-designed mimics of natural processes like photosynthesis. Mathematicians need to ask their scientific colleagues about these incredible ongoing research areas.

There is one final comment I want to make. Namely, to educate and prepare students to work in such a diverse, multi-disciplinary, and rapidly evolving area as nanoscience and technology, clearly there have to be changes in our educational systems. More math and science in K-12 is a given. And at the university level, the definition of a well-rounded, liberal arts education has to include scientific literacy. Especially in mathematics graduate programs, but in all scientific disciplines, students have to be encouraged, if not required, to take a battery of courses outside the field of their Ph.D. The emerging era of nanoscience and nanotechnology, with allied life science revolutions, will possibly contribute to a new Renaissance in developed society, one in which I surmise the role of mathematics will surpass that in the original Renaissance. I sincerely hope that the mathematics culture in this country rises to the challenges presented by the nanoscience and nanotechnology era in an unprecedented manner.

References

Adalsteinsson, D. and J. Sethian. 1997. An overview of level set methods for etching, deposition, and lithography development. *IEEE Transactions on Semiconductor Devices*. 10(1), 167-184.

Arias, T.A. and Sohrab Ismail-Beigi. 2000. Novel algebraic formulation of density-functional calculation. *Computer Physics Communications*. In press.

Baschnagel, J., K. Binder, K. Kremer et al. 2000. Bridging the gap between atomistic and coarse-grained models of polymers: status and perspectives. *Advances in Polymer Science*. Volume 152. Springer-Verlag.

Brandt, A. 2000. *Multiscale Scientific Computation: 2000*. Weizmann Institute preprint. October.

Caflisch, R.E., W. E, M.F. Gyure, B. Merriman, and C. Ratsch. 1999. Kinetic model for a step edge in epitaxial growth. *Phys. Rev. E*. Vol 59. 6879.

Cipra, Barry A. 2000. The best of the 20th century: editors name top 10 algorithms. *SIAM News*. vol. 33(4). Computing in Science and

Engineering, A.I.P and IEEE press, also covered by http://www.siam.org/siamnews/05-00/topten.pdf, Jan/Feb.

Colbert, D.T. and R.E. Smalley. 1999. Fullerene nanotubes for molecular electronics. *Trends in Biotechnology.* Vol 17. 46-50.

Dangl, J.L., 1998, Plants just say NO to pathogens, *Nature.* Vol. 394, 525-526.

Darden, T., L. Perera, L. Li, L. Pedersen, 1999, New tricks for modelers from the crystallization toolkit: The particle mesh Ewald algorithm and its use in nucleic acid simulations, *Structure.* Vol. 7, R55-R60.

de Gennes, P.G. 1974. *The Physics of Liquid Crystals.* Oxford University Press.

DeSimone, J.M., E.T. Samulski, et al. 2000. Critical micelle density for the self-assembly of block copolymer surfactants in supercritical carbon dioxide. *Langmuir.* Vol.16. 416-421.

Doi, M. and S.F. Edwards. 1986. *The Theory of Polymer Dynamics.* Oxford University Press (Clarendon), London-New York.

Donald, A.M. and A.H. Windle. 1992. Liquid crystalline polymers. *Cambridge Solid State Science Series.* Cambridge University Press.

Dym, H. and H.P. McKean,. 1972. *Fourier Series and Integrals.* Academic Press.

Fermi, E., J. Pasta, and S. Ulam. 1974. Studies of nonlinear problems, I. *Los Alamos Report.* LA1940. 1955.

Forest, M.G., Q. Wang, and H. Zhou. 2000a. Exact banded patterns from a Doi-Marrucci-Greco model of nematic liquid crystal polymers. *Physical Review E.* Vol. 61(6). 6665-6672.

Forest, M.G., Q. Wang, and H. Zhou. 2000b. Homogeneous pattern selection and director instabilities of nematic liquid crystal polymers induced by elongational flows. *Physics of Fluids.* Vol. 12(3). 490-498.

Fourier, J. 1878. *The Analytical Theory of Heat.* translated by A. Freeman. Cambridge University Press. London and New York.

Goldbeck-Wood, G. and A.H. Windle. 1999. Lattice modeling of nematodynamics. *Rheol. Acta.* Vol. 38. 548-561.

Grosberg, A.Y. and A.R. Khokhlov. 1994. *Statistical Physics of Macromolecules,* AIP Series in Polymers and Complex Materials, AIP Press, New York.

Hadjiconstantinou, N.G. 1999. Combining atomistic and continuum simulations of contact-line motion. *Physical Review E.* Vol. 59(2).2, 2475-2478.

Kruskal, M., and N. Zabusky. 1965. Interaction of solitons in a collisionless plasma and the recurrence of initial states, *Phys. Rev. Lett.* Vol. 15. 240-243.

Kupferman, R., M. Kawaguchi, and M.M. Denn. 2000. Emergence of structure in models of liquid crystalline polymers with elastic coupling. *J. Non-Newtonian Fluid Mech.* Vol. 91. 255-271.

Larson, R.G. 1999. *The Structure and Rheology of Complex Fluids.* Oxford University Press.

Lax, Peter D. 1989. The flowering of applied mathematics in America. in *A Century of Mathematics in America-Part II.* edited by Peter Duren et al. American Mathematical Society. 455-466.

Lee, M.S. and M. Head-Gordon. 2000. Absolute and relative energies from polarized atomic orbital self-consistent field calculations and a second order convergence with size and composition of the secondary basis. *Computers and Chemistry.* Vol. 24(3,4). 295-301.

Lipowsky, R. 1991. The conformation of membranes. *Nature.* Vol. 349. 475.

Michalet, X. and D. Bensimon. 1988. Vesicles of toroidal topology: observed morphology and shape transformations. *J. Phys. II.* France. Vol. 5. 598.

Nauta, K. and R.E. Miller. 1999. Non-equilibrium self-assembly of long chains of polar molecules in superfluid helium. *Science.* Vol. 283. 1895-1897.

Parr, R.G. and W. Yang. 1989. *Density-Functional Theory of Atoms and Molecules.* Oxford University Press. Oxford.

Pierce, N.A. and M.B. Giles. 2000. Adjoint reconstruction of superconvergent functionals from PDE approximations. *SIAM Rev.* Vol. 42(2). 247-264.

Senturia, S., N. Aluru, and J. White. 1997. Simulating the behavior of MEMS devices: computational methods and needs. *IEEE Computational Science and Engineering.* Vol. 4(1). 30-43.

Smith, D., H. Babcock, and S. Chu. 1999. Single-polymer dynamics in steady shear flow. *Science.* Vol. 283. 1724-1727. March.

Tadmor, E., R. Phillips, and M. Ortiz. 2000. Hierarchical modeling in the mechanics of materials. *Int. J. Solids Struct.* Vol. 37. 379.

Thomas, E.L., D.M. Anderson, C.S. Henkee, and D. Hoffman. 1988. Periodic area-minimizing surfaces in block copolymers. *Nature.* Vol. 334. 598.

Tsuji, T. and A. Rey. 1998. Orientation mode selection mechanisms for sheared nematic liquid crystalline materials. *Phys. Rev. E.* Vol. 57(5). 5609-5625.

Vaia, R., D. Tomlin, M. Schulte, and T. Bunning. 2000. Two-phase nanoscale morphology of polymer/LC composites. *Polymer.* In press.

Voter, A.F. 1997. Hyperdynamics: accelerated molecular dynamics of infrequent events. *Phys. Rev. Lett.* Vol. 78. 3908.

Warner, M. and E. M. Terentjev. 1996. Nematic elastomers: a new state of matter? *Progress in Polymer Science.* Vol. 21(5). 853.

Warner, M., E. M. Terentjev, R. B. Meyer, and Y. Mao. 2000. Untwisting of a cholesteric elastomer by a mechanical field. *Phys. Rev. Lett.* Vol. 85(11). 2320.

E, Weinan. 2000. Private communication.

Acknowledgements

I wish to acknowledge contributions from numerous colleagues who have made significant contributions to this essay: Russ Caflisch, David Cai, Roberto Camassa, Nick Ercolani, Bob Kohn, Dave, Ken, and Rich McLaughlin, Graeme Milton, Michael Rubinstein, and Sean Washburn.

IMPLICATIONS OF NANOTECHNOLOGY FOR THE WORKFORCE

S.J. Fonash, Penn State University

Introduction

Nanotechnology, which grew out of many fields, is becoming a very broadly based and very broadly utilized technology. It has now become the meeting ground of engineering, biology, physics, medicine, and chemistry. Nanotechnology is being used in all these fields and it, in turn, is utilizing elements of all these pure and applied sciences. The workforce carrying out nanotechnology manufacturing is composed of engineers and technicians. Each group must have a broad science base to function in this workforce and to take full advantage of the career opportunities it provides. The preparation of these two components of the nanotechnology workforce faces a number of problems. Some of these are common to both the education of the engineering workforce and the technician workforce. Some are specific to the education of each workforce component.

Engineering Workforce

The implications of nanotechnology for the four-year-degree engineering workforce are that engineers in this field will have to have a broad background encompassing an understanding of the principles of biology, physics, and chemistry as well as encompassing the engineering principles of design, process control and yield. The sciences provide the tools of nanofabrication and they also dictate the rules of the nanoscale world. Biology is needed for two reasons: manufacturing will increasingly mimic biological systems assembly and manufacturing will increasingly be fabricating systems for bio-medical applications. Physics is needed because the nanoscale is the world of probability wavefunctions, quantum mechanical tunneling, and atomic force probes. Chemistry is critical because it provides the tools for tailoring molecules, functionalizing surfaces and "hooking" everything together.

Unfortunately our engineering education system is not geared, for the most part, to teaching a unified approach to understanding and using science. At the secondary level, teachers, counselors, and administrators, for the most part, do not recognize the coming impact of nanotechnology and certainly do not know its educational demands. The approach of the secondary level continues to be a compartmentalized treatment of knowledge with the life sciences, physics, and chemistry each in its own box. In the secondary school world, students, parents, and teachers are familiar with terms such as information technology, the Internet and software, but there appears to be little awareness of the driver behind all this — nanoscience and nanotechnology.

At the college level increasingly fewer of these students from our secondary educational system find their way into the four-year degree engineering fields that deal with the development and manufacturing of "things". Secondary school students have grown up with computers at every turn and are at home with the concept of moving information around electronically. They never were exposed to nanotechnology manufacturing and are not at home with the concept of moving molecules around to build things. These students usually have no concept about the driver technology behind the computer screens or the Internet they use. Consequently, it is not surprising that at the college level fewer students are going into the engineering fields that create "things" and contribute to this driver technology.

For those students who do enter four-year degree college engineering programs, the educational approach awaiting them is one which focuses students and very infrequently tries to broaden their science education. More startling is the fact that the trend nationally is for four-year engineering programs to contain a diminishing exposure to science. For example, today's four-year degree electrical engineering students are not exposed to biology in college. Most are not exposed to any college level chemistry and the trend is for these students to receive less and less modern physics including quantum mechanics, which reigns supreme at the nano-scale. The engineering component of the nanotechnology workforce needs to know the tools available from, and the rules imposed by, the life sciences, physics, and chemistry but four-year degree engineering education is moving away from such a broad-exposure experience. Even the Accreditation Board for Engineering and Technology (ABET), which accredits four-year engineering degree programs, imposes criteria and regulations that work against a broad science exposure. There are some undergraduate engineering degree programs such as the engineering science program at Penn State and engineering science and engineering physics programs elsewhere that have an educational philosophy that fosters the broad science base needed in nanotechnology. However, such programs are few in number.

What can engineering educators do so that our colleges and universities can work to overcome these entrenched educational systems and trends that work against the broad science-based exposure needed by the nanotechnology engineering workforce? There are several possible solutions:

- More and broader exposure to science in undergraduate engineering programs.

- More reliance on waiting until graduate school to develop a deeper, broader science education.

- More reliance on post-graduation education of non-engineering biology, physics, and chemistry majors to add these four-year degree graduates to the engineering workforce

The first of these solutions will be difficult to implement especially in the near future. The climate today in engineering education is one where technology and specialization are stressed in undergraduate education.

Professional societies and ABET, which are more in tune with traditional technologies and traditional industries, will have to eschew specialized education and turn to generalization and a strong science background. This will be slow to happen at best.

The second of these solutions probably will continue to occur and expand. It is what happens now — engineers broaden and deepen their science background in graduate school. The difficulty with this solution lies in the fact that the growth in the number of Americans going to graduate school has, at best, leveled off in recent years. The result of this is that having a strong nanotechnology engineering workforce will require more reliance on attracting engineers from other countries.

The third solution depends on utilizing the background of the students who graduate each year with four-year degrees in biology, physics, and chemistry. If the "bridging courses" covering the engineering practice of nanotechnology development and manufacture were available in the engineering schools for these four-year degree science graduates, members of this group of students could enter the engineering nanotechnology workforce with one or two semesters of post-graduate education.

Technician Workforce

Two-year degree programs must be developed that create a nanofabrication technician workforce. Such a technician workforce is a requirement to have viable nanofabrication manufacturing. Developing this workforce faces many of the same problems as does developing the four-year degree engineering workforce. The common problems are centered on the secondary school system, which, as noted, does not yet understand the impact of nanotechnology and, in any case, does not foster a broad integrating approach to science education. In addition, developing a technician workforce faces a unique problem from the secondary education system: secondary schools tend to discourage the attendance at two-year degree colleges. Generally speaking, the secondary school environment seems to be almost exclusively focused on directing students to four-year degree programs even if this results in more people with multiple years of aimless academic wondering and four-year college graduates with unmarketable skills. Further, the vocational technology high schools of the secondary school systems, which could be an excellent resource for preparing students for entry into two-year degree college programs in nanotechnology, often concentrate on teaching activities such as computer repair and cosmetology. To summarize: to develop the nanofabrication technician workforce that will be needed by our society, our secondary school educational system must

- Understand what nanotechnology is and understand its unique position as a "driver technology".

- Provide secondary school students with a broader background in science.

- Revitalize vo-tech schools so they can prepare students for a nanofabrication technician education in two-year degree college programs.

- Encourage students to consider attending a two-year degree college for technical training.

- Foster the realization that a two-year degree allows one to enter the workforce quickly. This allows individuals to meaningfully assess careers and to then continue on to complete a four-year degree, when true interest and commitment develop.

From the point of view of our two-year colleges, creating a nanotechnology technician workforce is very demanding. The two-year colleges must insure their graduates have a broad base in science and technology and in the various applications of nanotechnology. It is only this type of education that can produce nanotechnology technicians that can move from industry to industry as the job opportunities shift back and forth among nanotechnology fields. Making sure that the nanotechnology workforce has the training and skills to apply nanotechnology to industries as diverse as biomedical applications, MEMs, pharmaceuticals, opto-electronics, information storage, and of course, microelectronics is critical. It is critical because all these industries will need this workforce. It is critical because educators must create workforce-training programs that give technicians the self-assurance and freedom to use their nanotechnology skills in the field with the best opportunities. An effort is needed here too for another reason: the semiconductor industry's history has caused those considering the nanotechnology technician workforce to fear a "boom or bust" job environment. To overcome this fear and to attract more individuals, a nanotechnology workforce must be created that is science-grounded and skill-based so that it is not trapped in one industry. The two-year degree colleges must develop the curricula needed to give this generalist, nanotechnology technician education.

Analogous to nursing programs, nanotechnology technician education must expose students to the "operating rooms" of nanotechnology; i.e., to cleanroom suites with their deposition, etching, materials modification, and lithography functions. Two-year degree colleges cannot afford to establish and maintain these state-of-the-art facilities just as they cannot establish and maintain hospital operating rooms for nursing programs. As with nursing programs, the answer will have to be resource sharing; i.e. those major research universities that have such facilities must share their nanofabrication laboratories with two-year colleges for nanofabrication technician workforce development.

These needs for a nanofabrication technician work force mean that our two-year colleges must

- Create curricula that produce technicians who are not captive to one narrow segment of industry but who have broad background in science and nanotechnology processing.

- Give students the tools and confidence to be able to move as they wish, among such diverse industries as pharmaceuticals, MEMs, information storage, opto-electronics, biomedical applications, and microelectronics.

- Train students in state-of-the-art nanofabrication facilities.

- Resource share with those major research universities that have state-of the art nanofabrication facilities to give two-year degree students hands-on experience in state-of the art environments.

Facilities

Since nanofabrication facilities with a sustaining staff are difficult to create and expensive to maintain, many engineering schools, in addition to the two-year degree colleges, cannot offer students any exposure to the practice of nanofabrication. It could be argued that this exposure will take place in industry and that the role of four-year engineering programs preparing an engineering workforce for nanofabrication is to insure the understanding of the tools and limits provided by science and the understanding of the manufacturing and yield principles of engineering. However, the counter argument is that many of these engineering students will never even see a nanofabrication facility in their four-year programs. Consequently many may lose their enthusiasm due to a lack of hands-on experience; surely many will never even go into the field because they have never seen the practice of "engineering at the atomic scale." The National Science Foundation is addressing the problem in a Research Experience for Undergraduates (REU) program that gives undergraduate students access to the nanofabrication facilities of NSF's National Nanofabrication Users Network (NNUN). However, this program is relatively small and does not target engineering students. More programs giving more engineering undergraduate students access to state-of-the art nanofacilities will be needed for development of an engineering workforce skilled in nanofabrication.

This problem of facilities access is even more critical for developing the technician workforce for nanofabrication. The technician workforce must be hands-on since manufacturing is its primary function. Yet the two-year colleges cannot support nanofabrication facilities, as noted earlier. Pennsylvania has developed a unique approach to this workforce development issue: the Pennsylvania Nanofabrication Manufacturing Technology (NMT) Partnership. This Partnership joins the state, the Penn State site of the NSF National Nanofabrication Users Network, industry, and two-year colleges together in a statewide NMT technician workforce development program. It is a resource-sharing partnership in which Penn State's Nanofabrication Facility is shared with the two-year colleges. The PSU Nanofabrication Facility offers a one-semester "capstone experience" for students at two-year institutions in this program. This "capstone experience" consists of six hands-on courses in nanofabrication taught at the Penn State Nanofabrication Facility each semester for two-year colleges. These courses belong to the two-year colleges, they use them in their nanotechnology technician training, they give credits for these courses, and they give two-year NMT degrees.

The value of a nanofabrication technician workforce with hands-on training can be measured by the salaries commanded by students that have completed the Pennsylvania NMT Partnership "capstone experience" courses. The starting salaries for these two-year degree nanofabrication manufacturing technology graduates who have accepted employment offers are in the $40,000 to $46,000 per year range with students receiving up to seven offers from industries in Pennsylvania.

Summary

Nanoscience and nanotechnology are the meeting ground of biology, engineering, medicine, physics, and chemistry. As a result, development and manufacturing at the nano-scale will be increasingly involved with components of all these pure and applied sciences. Because this field is becoming so broadly based and so broadly utilized, practitioners will no longer be limited to the microelectronics industry as they once were but, instead, will be free to choose jobs and follow career paths in a variety of fields such as bio-medical applications, MEMs, microfluidics, opto-electronics, information storage, pharmaceuticals, and, of course, microelectronics.

Students considering careers in nanotechnology must be made aware of the broad educational requirements of the nanotechnology workforce. First, however, educators must be made aware of what is required to prepare this workforce. "Getting the word out" and preparing middle and high school students properly will require informed secondary school teachers, counselors, and administrators. At the next step, colleges and universities must create an engineering and technical nanofabrication workforce with the broad science and technology background needed to carry out nanotechnology manufacturing. Developing a broadly trained and educated workforce presents a severe challenge to our four-year degree and two-year degree educational systems, which favor compartmentalized learning. Also this nanofabrication workforce will have to be exposed to state-of-the-art nanofabrication facilities in the course of its training and education. Given the cost of creating and sustaining such facilities, incorporating their use into nanotechnology workforce development across the nation presents a considerable challenge.

SOCIETAL IMPACTS OF NANOTECHNOLOGY IN EDUCATION AND MEDICINE

V. Vogel, University of Washington

Educating the Next Generation

In most disciplines, education has progressed by first laying a foundation and then building pyramids of knowledge step by step. This approach to education has resulted in a highly specialized workforce. It has promoted enhanced departmentalization in academia, each field imprinting its own way of thinking on its scholars and evolving its own languages and acronyms. The level of specialization has progressed over the decades, deepening the trenches between disciplines to such an extreme that publications often became incomprehensible to scholars outside the field. Such a divergence in science makes it difficult for one discipline to capitalize on the advances of another. The discovery of new analytical tools to visualize and manipulate single atoms, however, have marked a turning point from divergence to convergence within the scientific community. With atomic force microscopy, optical tweezers and single molecule spectroscopy at hand, scientists and engineers in a variety of fields have started to explore the nanoscale world. The frontiers of many disciplines, including physical sciences, biosciences and engineering, have started to converge at the nanoscale, and nanotechnology has begun to thrive from this interdisciplinary cross-fertilization. But moving beyond the confines of a secure research environment into relatively unknown territory is an academically risky endeavor. It will, however, be an inevitable step for many even in this generation of scientists, and even more so for future ones. Pioneering into new areas therefore must be facilitated by increasing funding for interdisciplinary research programs, co-advising of students across departmental lines, and by enhancing communication across the disciplines through newly tailored workshops and summer schools.

What impact will this new zeitgeist have on how the next generation of graduate students will be educated? Since nanotechnology encompasses a variety of disciplines, including the physical sciences, engineering and biomedicine, an educational system focusing on single disciplines will not provide adequate training. Yet, it will remain essential to the vitality of this emerging field that challenges be tackled from different perspectives, by people who communicate well but have different mind-sets and expertise. Therefore, introducing new degree programs in nanotechnology that provide a shallow overview of many disciplines, but none in sufficient depth to make major contributions, will not give our students the training they need to meet the future challenges. Recognizing that the convergence of technology has to be reflected in education, we have just introduced a new Ph.D. program in nanotechnology at the University of Washington, Seattle. Funded through a NSF-IGERT award in 2000, our graduate students will receive an in-depth education in one of nine participating home departments, yet gain early exposure to the other disciplines through additional course work, joint seminars, and by being co-mentored across departmental lines. The goal is that once they leave the program, they will still think like physicists, chemists, biologists or engineers, yet they will have developed enough awareness of other disciplines to capitalize on progress in those other fields. They will have the ability to effectively communicate and to lead interdisciplinary research teams in both industry and academia.

Creating an environment in which students can obtain an interdisciplinary education is not the only challenge. Today, major technological breakthroughs occur within just a few years, which approaches the typical time scale for completing a Ph.D. thesis. For the first time in history, graduate students may soon face a situation where the technical skills that they learned at the beginning of graduate school become obsolete when they graduate. This implies that future graduate students may not be hired any longer for their specific technical expertise but for more general talents, including their ability to think, how fast they learn, and how they find and disseminate information to solve problems. An exponential growth of technology will also select for those individuals who have an aptitude to constantly learn and adjust. If this is to become the new paradigm for hiring decisions, universities have to rethink how to best prepare their graduate students for this new environment.

The astonishing rapidity with which the Internet is changing the world has made it clear that education and technology have to be integrated from elementary school through to the graduate level. Playing prepares children to solve real-life problems, and they will play with whatever technology is available. Children will thus grow up used to exploiting the latest technology to learn and will expect to keep doing so in their classrooms. This will increasingly pose a challenge to teachers to stay current with developments in their field and to integrate new tools and technology into their teaching. It cannot be left to the teachers to face this struggle alone. Schools and society have to think how to best help their teachers to remain effective educators, for example, by hiring technical support staff to help teachers in the classroom. Research is also needed to assess how interactive games and teaching modules affect and change learning behavior.

Preparing for Rapid Advances in Medicine: the Fine Line Between Reality and Science Fiction

Science fiction has played a powerful role in creating excitement and fascination for nanotechnology in society. It envisions fictitious scenarios of both the meritorious and the dark side of nanotechnology. The public perception of the power of nanotechnology in medicine is still dominated by scenarios described in movies and science fiction literature. In the movie "Fantastic Voyage" from the 1960s, for example, macroscopic design principles were scaled down to the nanoscale to create nanobots — nanoscale robots — that travel through the blood stream equipped with intelligently controlled arms and sensors. In Star Trek, the blind wear goggles that transmit visual images directly to the brain; and with the command "Energize" people are transported across space, disassembled and reassembled in an instant. Robots are often indistinguishable from humans, able to rationalize, argue and perform intelligent tasks. And the endless dream of immortality is portrayed by some to become reality once society fully exploits its new nanotech toys.

What remains the fabric of dreams and what will become reality? There have certainly been stunning examples of shortsightedness in predicting the future, even from leading figures in science and technology. This, however, does not mean that scientists should shy away from helping society plan for its immediate future. In fact, scientists must play a vital role in drawing the line between realistic predictions and futuristic dreams, to prevent the public podium from remaining occupied by scientifically unsubstantiated optimists or worriers. An attempt is made below to define that fine line between science fiction and reality, addressing the most publicly disputed nanotech visions in medicine. The focus is on short (2-3 years), medium (5-15 years), and long-term (over 20 years) perspectives.

Short-term perspective: Nanobiotechnology is a rapidly advancing frontier which has already catalyzed an explosion of entirely new industries in health care, medicine, food and nutrition, environmental management, chemical synthesis and agriculture. Advances in nanoanalytical tools and engineered nanoscale systems are converging with the rapid progress made in genomics, combinatorial chemistry, high throughput screening and sequencing, drug discovery, microfluidics and bioinformatics. Nanobiotechnology will also bring tremendous advances in the early detection of diseases and their treatments, and in our fundamental understanding of pathogenic pathways. New nanoanalytical tools are pushing detection limits down to the single molecule level, which is scientifically a huge success but could be a potential headache to regulators. Ultrasensitive detection of toxins and pollutants will alarm the public. The public will demand that accurate and scientifically defendable health risk thresholds be defined, which will often be a nontrivial task. Furthermore, new DNA chip technology is at the verge of being able to probe for genetic predispositions at affordable cost. The ethical implications of predictive technology to assess the mental and physical health of patients are far-reaching and are being heavily discussed in conjunction with the human genome project.

Medium-term perspective: The shortage of organ transplants is already a major problem that is only likely to worsen as the population rapidly ages. Efforts are thus underway to develop synthetic organs. Advances in nanoengineered materials combined with a molecular scale understanding of wound healing and tissue repair processes will be key to integrating engineered biomaterials into biological tissue and to engineering tissue and organs that will take over at least some vital functions of failing organs. The first artificial skin has received Food and Drug Administration (FDA) approval in the United States. Society will also see major advances in treating the loss or partial loss of auditory, visual, and sensory functions through the introduction of novel micro and nanoengineered electronic devices. Some of these may be hidden in goggles, behind the ear, or in modern interactive clothing; others may be directly implanted. Cochlear implants that record a broad spectrum of frequencies are already available.

Visual image-enhancement or processing implants may be feasible within a decade.

Long-term perspective: Projecting exactly how long it will take until electronic devices will be implanted into the brain to enhance or compensate for lost brain function is more difficult, but that time will eventually come. Major progress has already been made in recording from single neurons and their stimulation, and culturing nerve cells on microelectronic devices (Bai, Wise, and Anderson 2000). It seems likely that technology will be able to control at least a few simple brain functions by the use of brain implants. While this will allow for a tremendous improvement in the life quality for some patients, various ethical issues will have to be addressed. It will be the task of the FDA to regulate experimentation, fabrication and usage of brain implants, but the implications of such implants reach much further than safety and ethics. For example, current law considers whether crimes are conducted under the influence of drugs. In the future, one may have to ask whether a person's state of mind has been impacted by the influence of externally addressable brain implants and if so, who is responsible for their actions.

Science fiction rather than reality: The popular press has often aired heated discussions by the author Ray Kurzweil and others about the idea that it will soon be possible to scan the human brain and essentially transfer its neural activity to a computer designed to simulate billions and billions of human neurons (Kurzweil 1999). This fantastic thought is based on a series of assumptions, some of which are reasonable extrapolations of future technological abilities. Others, however, completely neglect how little is still known about how the mind works. Imaging technology may indeed reach microscopic resolution, which may reveal individual synaptic contacts between nerve cells. If Moore's law can be extrapolated, computers will achieve the memory capacity and computing speed of the human brain by around the year 2020. Computer experts were therefore quick to postulate that copying the 3D neural circuitry of the human brain would become possible with these powerful computers and advanced imaging technologies. Once this is achieved, they claim, it will be possible to simulate first the brain and its function and eventually the state of the human mind, complete with its memories, emotions and creativity. But it is important to remember that these nightmarish scenarios are put forward without any real biological understanding of the brain. For example, these scenarios rely on the assumption that the brain is nothing more than a hard-wired neural network, and that knowledge of the 3D brain architecture would be sufficient to assess its functional states. This may be the case for nematodes — the worm C. Elegans has a nervous system consisting of 302 neurons whose connections are all known (White 1986). But the brains of higher vertebrates have fundamentally different system architectures than computers. Furthermore, single neurons are highly nonlinear systems. Single neurons in the cerebrum can make more than ten thousand connections to other nerve cells. The picture gets even more complex with the recent findings that higher vertebrate brains show plasticity. Plasticity is the ability of a system to change its structure and/or function in response to injury, the environment and/or other changing conditions (for further readings see Jacobs et al. 2000; Malinow, Mainen, and Hayashi 2000; Poldrack 2000; Simos et al. 2000; Tramontin and Brenowitz 2000). Given this complexity of the brain, a scan of the brain will not allow a read-out of the human brain's mind nor its memory. A century ago, society was embroiled in an almost parallel controversy as to whether the future was completely deterministic and calculable based on Newtonian mechanics. It took the discovery of quantum mechanics to defy the notion that our future is predictable.

Nanobots are often on top of the list of nanotechnological creations that cause deep concern to the public. Eric Drexler and followers postulate that it will soon be possible to create nanoscale, addressable robots that have the ability to move in space, recognize the environment and self-replicate (see e.g., Stix 1996 for a critical review). Will it indeed be possible to create another form of life at the nanoscale? When it comes to the engineering of nanoscale machinery, nature is still far superior in its ability to integrate synergistically operating nanoscale systems of high complexity. Yet, even nature has not been able to engineer nanoscale creatures that combine all of the above-mentioned attributes of nanobots. Viruses are amazing nanoscale systems, but even they do not have the finesse of the hypothetical nanobots. Viruses are able to move and they contain the genetic blueprint of themselves, yet they are not capable of self-replication. Since they depend on the replication system and protein synthesis machinery of much larger organisms, namely micro-scale cells, they do not meet the definition of a self-replicating system. While mankind is equipped with increasingly powerful tools to manipulate living systems, we are not at the verge of creating herds of synthetic self-replicating nanobots that will run out of control and threaten our lives. Future man-made nanosystems will certainly be able to perform a variety of functions, but a robot that is proficient in all three functions — movement in space, recognition of a chemically complex environment and self-replication — will remain the fabric of dreams.

References

Bai, Q., K.D. Wise, and D.J. Anderson. 2000. A high-yield microassembly structure for three-dimensional microelectrode arrays. *IEEE Trans. Biomed Eng.* 47(3): p. 281-9.

Jacobs, K.M., et al. 2000. Postlesional epilepsy: the ultimate brain plasticity. *Epilepsia.* 41(Suppl 6): p. S153-61.

Kurzweil, R. 1999. *The Age of Spiritual Machines.* Penguin Books.

Malinow, R., Z.F. Mainen, and Y. Hayashi. 2000. LTP mechanisms: from silence to four-lane traffic. *Curr. Opin. Neurobiol.* 10(3): p. 352-7.

Naughton, G. 1999. The Advanced Tissue Sciences story. *Sci. Am.* 280(4): p. 84-5.

Poldrack, R.A. 2000. Imaging brain plasticity: conceptual and methodological issues — a theoretical review. *Neuroimage.* 12(1): p. 1-13.

Simos, P.G., et al. 2000. Insights into brain function and neural plasticity using magnetic source imaging. *J. Clin. Neurophysiol.* 17(2): p. 143-62.

Stix, G. 1996. Trends in nanotechnology: waiting for breakthroughs. *Sci. Am.* April.

Tramontin, A.D. and E.A. Brenowitz. 2000. Seasonal plasticity in the adult brain. *Trends Neurosci.* 23(6): p. 251-8.

White, J.G. 1986. *Philos. Trans. R. Soc. Lond. B Biol. Sci.* 314: pp. 1-340.

TECHNOLOGICAL AND EDUCATIONAL IMPLICATIONS OF NANOTECHNOLOGY — INFRASTRUCTURE AND EDUCATIONAL NEEDS

J.L. Merz, University of Notre Dame

Introduction

This paper leads off with a brief review of the National Nanotechnology Initiative, to set the stage for what follows. The paper deals primarily with both research infrastructure and graduate education (which can be viewed as another infrastructure component). This subject has less to do with society as a whole, and more to do with the *culture* of how we *do* science, and how we *teach* science. The discussion leads into some reflections on the relative funding for the physical sciences vis-à-vis the life sciences. The paper then concludes with a few comments on some of the ethical implications that may arise as the country pursues the National Nanotechnology Initiative.

The National Nanotechnology Initiative

The President proposes spending approximately half a billion dollars on the National Nanotechnology Initiative (NNI) (NSET 2000), which champions fundamental research directed at developing a long-term nanotechnology that would revolutionize many areas of human activity as we know them today. Nanotechnology is expected to have a major impact on information processing and technology, communications, biomedicine and health, transportation, and the environment. Specifically the President called for an 84% increase over the $270 million currently spent on nanotechnology at all agencies, to a total of $497 million, with specific budget increases recommended for each of the agencies. Notable is a 124% increase recommended for the National Science Foundation (NSF), currently the largest supporter of nanotechnology at $97 million in FY 2000. Although the budget has not been finalized at the time of writing of this article, the Senate has approved legislation providing an overall increase of $529 million for the NSF, which represents a 13.6% increase, the largest ever received by the Foundation. It is expected that the House will pass this bill without amendment, clearing the way for the President's signature. The hoped-for NSF increase of 124% in the area of nanoscience was reduced to 33%, providing an additional $52.7 million to the current $150 million for the Nanoscale Science and Engineering initiative.

NNI was developed by the Interagency Working Group on Nanoscience, Engineering, and Technology (IWGN) which was formed by the National Science and Technology Council. IWGN convened a workshop in January 1999, and published its report from that workshop in September 1999 (Roco et al. 1999). In this report a series of recommendations are offered for expenditures in five different areas, listed below. The President's FY 2001 budget embraces all five of these areas, with percentage allotments of the $497 million as shown.

- Fundamental Research   39%

- Grand Challenges   22%

- Centers and Networks of Excellence 16%

- Research Infrastructure   17%

- Workforce Implications   6%

These five initiatives represent an effective strategy to realize the ambitious goals of the NNI. They appropriately emphasize fundamental,

187

interdisciplinary, long-term research in a variety of contexts: single investigator or small teams from different disciplines (Fundamental Research), larger research teams forming centers that may be located at a single university or distributed among a number of locations (Centers and Networks of Excellence), interdisciplinary research and education teams (Grand Challenges) that may involve the Centers and Networks of Excellence, the development of new tools for carrying out the research (Research Infrastructure), and the support of education and training of the future workforce, including the creation of graduate student fellowships that are not tied to a specific discipline (Workforce Implications). These recommendations, and the President's promotion of them, represent an unprecedented opportunity for researchers in this important field, moving significantly in the directions needed. Most of the pieces are there.

However, some critical elements are missing. All of the elements above are part of a research paradigm that has been developing over the last 15-20 years. The importance of single investigators and small teams has never been questioned. On the other hand, the emergence of research centers that involve larger numbers of faculty at a single university or embracing several campuses, with significant educational outreach components, has resulted from NSF's formation of a variety of research centers, such as Engineering Research Centers (ERCs), Science and Technology Centers (STCs), Materials Research Science and Engineering Centers (MRSECs), etc. The development of new measurement capabilities has been fostered by NSF and DOD instrumentation grants, and a few examples of equipment user facilities exist, the most relevant of which is the National Nanofabrication Users Network (NNUN). So many of the elements of NNI are already in practice; the President's initiative adds considerable resources to them, and provides researchers with an unparalleled opportunity. The *missing* pieces, however, are essential infrastructure resources, both capital equipment resources and human intellectual resources, that go well beyond the recommendations of the workshop report.

The Need for Capital Equipment Resources

Capital equipment facilities for the fabrication of nanostructures, the analysis of surfaces and interfaces between nanoparticles and the material in which (or on which) they are located, and the characterization of the novel physical, electronic, thermal, and other properties anticipated for these nanostructures, are badly needed for *all* investigators in this initiative. In the old paradigm, individual investigators or members of research centers try to acquire the facilities needed by them through application to the various agency programs. This results in unnecessary duplication of resources on the one hand, and their fragmentation on the other. NNI proposes no radical steps to change this. What is needed is the availability of flexible, accessible, user-friendly, modular facilities for the fabrication, analysis, and characterization of structures to researchers all over the country. At least ten such facilities should be created, which researchers can access in much the same fashion as astronomers access telescopes and high energy physicists access accelerators. The facilities need to be staffed by technical experts who install and maintain state-of-the-art equipment and assist/train researchers in its use. Note that this is very different from the concept of ten centers or networks of excellence proposed in the NNI, for the centers of excellence carry out a research *program*, whereas the facilities proposed here would provide the *infrastructure* needed for each center's program. Such user facilities are mentioned only peripherally in NNI, and little money is allocated for them.

The National Nanofabrication Users Network (NNUN) mentioned above represents the first step in this direction, for it is truly a network of laboratories made available to outside users from other universities and from industry. However, the network is small, consisting of only five sites: Cornell, Stanford, UC Santa Barbara, Penn State and Howard University. More serious is the fact that NNUN confines its activities to fabrication, and contains no services for materials growth or for characterization. The national initiative could provide the resources to significantly expand NNUN both in size and in scope.

These user facilities need to be distributed geographically proximate to the major research laboratories, centers, and universities. It would probably be wise to specify a "theme" or focus for each facility, for example, nanoelectronics, composite structures, biological and biotechnological materials, energy and chemicals. Or, to further encourage the interdisciplinary nature of this field, the focus could be on the type of equipment, such as equipment for the synthesis of materials, ultra-high resolution scanning probes, or high-energy electron microscopes.

One may consider at least three models for the development of these user facilities. Combinations of two or three of these models may also be possible.

- *University-based.* The research universities are very suitable sites for these user facilities, but it is essential that the facilities be separate from faculty or research-center programmatic labs in use. There might be three such centers, each providing the laboratory equipment and other infrastructure needed for progress in nanoscience and technology research. Examples might be:

- East Coast Consortium, including, among others, Cornell, MIT, Harvard, Yale, North Carolina State, Georgia Tech.
- Midwest Consortium, involving such universities as Purdue, University of Illinois, University of Michigan, Ohio State, Penn State, Notre Dame.
- West Coast Consortium, including Berkeley, Stanford, UC Santa Barbara, UC San Diego, UT Austin.

- *National laboratory-based.* Many of the national laboratories would be ideal sites to develop the user facilities described here. However, it is essential that they be (a) *unrestricted* in terms of access, and (b) *separate* from the on-going programmatic laboratories already operating on these sites. Examples of suitable labs would be:

  - DOE labs: Sandia, Lawrence Berkeley National Laboratory, Argonne, etc.
  - DOD labs: Naval Research Laboratory, Air Force Labs, Army Labs.
  - NASA labs, such as the Jet Propulsion Laboratory
  - Department of Commerce labs: NIST, NREL

- *Industry-based.* There are a few cases of laboratories sponsored jointly by the federal government and those industries having a stake in this area of activity, with access extended to researchers both from universities and the sponsoring industries. There are "existence proofs" of this concept: Stanford's Center for Integrated Systems (CIS), and the Microelectronics Center of North Carolina (MCNC). Also, the Japanese MITI joint research laboratories are appropriate examples, where ten or more companies participate, send their best researchers, and work only on fundamental problems which are "pre-competitive" (and hence not serious patent/IPR risks) (Merz 1986).

Combinations of the university, national laboratory, and industry laboratories could be united into infrastructure centers on a regional level. Thus, the means to pursue this research would be made available to a region of the US, where travel and communications between research groups is feasible, with no duplication of equipment, which is often the case when neighboring universities submit competing proposals to a federal agency. The center would also help solve personnel infrastructure issues: technicians to operate and maintain the equipment, other support staff.

The Need for Human Intellectual Resources

NNI proposes a fellowship program for multidisciplinary studies, where the fellowship is not tied to a single discipline. This sounds good, but it is difficult to achieve, due to the discipline-oriented culture of universities. Section 11.2 of *Nanotechnology Research Directions* (Roco et al. 1999) describes this problem:

> Although change is occurring in universities in relatively rapid fashion, there still exist many elements in the culture of our research universities that discourage multidisciplinary research. Examples include the administrative autonomy of academic departments and colleges, the fact that many centers and institutes "compete" with departments in terms of contract and grant proposal submission, the difficulties of determining (particularly with respect to tenure and promotion decisions) the relative creative contributions of faculty to multi-authored publications, and the unfortunate disconnect between research and teaching that is too often the case.

Another serious problem has developed in recent years, due to the extremely attractive employment opportunities provided by software companies such as Microsoft, and, more recently, the rapid development of the Internet and the consequent emergence of entrepreneurial opportunities which promise huge payoffs. It is becoming increasingly difficult to attract the best graduate students, particularly domestic graduate students, to the physical sciences and engineering (physics, chemistry, math, electrical, chemical, and mechanical engineering) which are *essential* to nanotechnology. Instead, computer science departments are bursting at the seams, while the physical sciences must recruit students intensively, and frequently with little success. Approximately half of these students are international, and a large number of them return to their homelands after receiving their degrees, unlike the situation 10 or 20 years ago when most stayed in the United States.

The emphasis on graduate fellowships for domestic students called for in NNI should therefore be given a prominent place in the initiative, but two requirements should be placed on their award:

1. that these fellowships be reserved exclusively for students in the physical sciences and engineering as listed above, and

2. that they be made truly interdisciplinary by requiring that the student be assigned to an interdisciplinary *project* led by two or more faculty members, rather than to either an academic department, or to a single faculty member within a department.

The second of these requirements should force the universities to begin addressing their problems of promoting and rewarding interdisciplinary research.

The problem of shortages of human intellectual resources in the physical sciences has been eloquently addressed by Prof. Richard Smalley, Nobel Laureate from Rice University (Smalley 2000), who sees nanotechnology as a real solution:

> The physical sciences (chemistry, physics, materials science) need a big boost. Funding in these critical areas have (sic) been flat for many years. As a direct consequence a severe shortage of bright young faculty has developed in these areas, and few young American boys and girls are electing to go into these areas in graduate schools throughout the US. But this is precisely the area from which the nanotechnology revolution will come. Chemistry, physics, and materials research are at the core of nanotechnology. These are the fields that discovered the atom and understood its inner workings, and developed the science of combining them in precise structures, and developed tools with which these nanostructures are probed and visualized. These are the fields that are developing the requisite fundamental knowledge, and the computational algorithms to realistically predict behavior. As nanotechnology develops, the critical, core areas of physics and chemistry in our nation's universities will become much more intimately coupled to engineering, to industry and society as a whole (biotechnology is doing this now for large sections of the classic life sciences). This greater relevance and higher-level funding will attract American youth to these classic core fields of science as never before. Trained in the physics and chemistry of nanotechnology they can reasonably expect to get high paying, high technology jobs that are of great social significance.

The Need for Monetary Resources

If the 20th Century has been the century of the physical sciences, as has often been stated, and the 21st Century will be that of the life sciences, there may be serious detrimental funding implications for the physical sciences. This problem is sometimes described in terms of the ratio of NSF to NIH funding, which has been getting smaller. (This is clearly too narrow a description, as considerable fundamental research in the physical sciences is also funded by mission agencies located in departments such as the Departments of Defense, Energy, and Commerce, but the NSF/NIH ratio is a useful yardstick). The point has already been made above that the physical sciences are critical to success in the field of nanoscience. Further, most of the potential sites suggested for laboratory user facilities are physical science labs. In short, nanoscience and technology are deeply dependent on a strong supply of research talent in the physical sciences, and the laboratory infrastructure needed to sustain it.

The mutual interdependencies between the physical and life sciences is nowhere more apparent than in the diagnosis and treatment of cardiovascular disease. In recent years a revolution has occurred in the development of diagnostic equipment: coronary angiograms, acoustic doppler echocardiograms, carotid ultrasound tests, electron-beam-induced-X-ray catscans, radioactive isotope stress tests and catscans that could determine specific pathways of blood flow constriction. These diagnostic tests were all developed by physical scientists working in close collaboration with clinical and medical researchers: cathode ray tube phosphors and/or liquid crystal displays, acoustic sonograms and tomographic techniques, radioactive tracers, high-vacuum electron beam scanning techniques, and highly sophisticated programming and data analysis to provide critical quantitative information in real time. The great advances of medical diagnostics and treatment in recent years were made possible by physical science. The lesson is clear — we must find a way to support *both* physical and life sciences if we hope to eradicate disease and add to the quality of life.

Ethical Considerations

We are on the verge of the unprecedented availability of personal medical data. Section 8.3 of the *Nanotechnology Research Directions* (Roco et al. 1999) describes this problem:

> Integrated nanoscale sensors could monitor the condition of a living organism, the environment, or components of the nutrient supply, sampling a range of conditions with a high degree of sensitivity. With arrays of ultraminiaturized sensors that sample a range of chemicals or conditions, the confidence level and specificity of detection would be much greater than is now possible with separate macroscopic sensors. ... One can project that in the next century highly sophisticated, small, and inexpensive sensors employing nanotechnology will be available and used routinely in many parts of our lives.

The data derived from such sensors can be used for the highly desirable monitoring and treatment of a patient's condition, or for forms of manipulation that may be unwanted by the patient. Ethical theologians speak of "convergences" of unrelated events that converge to a totally unexpected (and sometimes undesired) result. A striking example of a desirable convergence in the physical sciences was the unrelated but nearly simultaneous development of semiconductor lasers and optical fibers during the early 1970s (Pollock 1995).[4] The room temperature semiconductor laser diode was developed jointly at laboratories such as Bell Labs, IBM, the University of Illinois, and the Ioffe Institute in St. Petersburg, Russia, building on the conceptual ideas of visionaries such as Z. Alferov, R. Hall, N. Holonyak, and H. Kroemer (two of whom recently won the Nobel Prize in Physics for this work). The development of low-loss optical fibers occurred at Corning Glass Works during the same period. The convergence of these unrelated research activities made possible the field of optical communication, with its lightweight and inexpensive cables to deliver extremely high bandwidth communications needing little regeneration and amplification.

More recently there have been convergences between the physical and biological sciences, which allow the development of nanodevices that will change the face of medical diagnosis and treatment.[5] For example, the development of tiny sensors is anticipated that could be placed in the human body through implantation or injection into the bloodstream. These sensors could measure the chemistry and biochemistry of the host, collecting unprecedented quantities of data, and might even be able to broadcast this information to remote receivers using wireless techniques. Considerable progress in this field is currently underway. To cite a few examples of the immense effort that is unfolding in this area, biocompatible sensors are being developed with potential for use *in vivo* (Chen et al. 1999), hybrid nano-electro-mechanical (NEMS) devices powered by biomolecular motors are being developed for application to biosensors and self-assembling, sub-cellular NEMS devices (Bachand and Montemagno 2000), a transcutaneous power source is being investigated to drive a totally implantable artificial heart (Matsuki et al. 1996), and highly luminescent semiconductor quantum dots have been coupled to biomolecules for use in ultra-sensitive biological detection (Chan and Nie 1998). Research is underway to power implantable devices remotely and to transmit information between them and external

---

[4] An excellent review of the development of the double heterostructure semiconductor laser is provided by H.C. Casey, Jr., and M.B. Panish (1978).

[5] Cf., for example, the article by S.C. Lee (Lee 1998), "The nanobiological strategy for construction of nanodevices," and other articles in *Biological molecules in nanotechnology: the convergence of biotechnology, polymer chemistry and materials science*, edited by S.C. Lee and L. Savage (IBC Press, Southborough, MA, 1998).

data stations (Dudenbostel et al. 1997, Von Arx and Najafi 1997, Matsuki et al. 1996). Extraordinary improvements in the miniaturization of computational architectures will be required to accompany these *in vivo* devices — one such example of this is the current research on quantum-dot cellular automata, a totally new paradigm of transistorless logic circuits of nanoscale dimensions (Porod et al. 1999).

At the same time that these rather incredible research advances are taking place, there are many who would like to take advantage of the resulting diagnostic information in ways that might not be beneficial to the patient. For example, those responsible for managing medical insurance would find these data extremely valuable, leading to a convergence which might result in more affordable medical treatment (an ethically good outcome), but, at the same time, in the uninsurability of the patient (a demonstrably bad outcome, at least from the patient's point of view). The development of these new technologies inexorably lead to fundamental questions such as "Who has access to this medical diagnostic information", "To what uses may it be applied", and, once cures are developed, "Who has access to extremely costly cures, such as for AIDS"?

The opportunity and the challenge that we face, as we begin our research and development of nanoscience and technology, is to ask these questions at an early stage of our work, so that serious moral dilemmas can be averted. And the time is short, as the progress in this field is expected to be extremely fast.

References

Bachand, G. and C.D. Montemagno. 2000. Constructing organic/inorganic NEMS devices powered by biomolecular motors. *Journal of Biomedical Microdevices.* 2, No.3, 179-184.

Casey, Jr., H.C., and M.B. Panish. 1978. *Heterostructure Lasers, Part A: Fundamental Principles.* (Academic Press. New York). pp. 1-9.

Chan, W.C.W. and S. Nie. 1998. Quantum dot bioconjugates for ultrasensitive nonisotopic detection. *Science.* 281, 2016 (25 September).

Chen, C.-Y., K. Ishihara, N. Nakabayashi, E. Tamiya, and I. Karube. 1999. Multifunctional biocompatible membrane and its application to fabricate a miniaturized glucose sensor with potential for use *in vivo. Journal of Biomedical Microdevices.* 1, No.2, 155-166.

Dudenbostel, D., K.-L. Krieger, C. Candler and R. Laur. 1997. A new passive CMOS telemetry chip to receive power and transmit data for a wide range of sensor applications. *Proceedings of Transducers '97, an*

International Conf. On Solid-State Sensors and Actuators, Chicago, June (published by the IEEE). p. 995.

Lee, S.C. 1998. The nanobiological strategy for construction of nanodevices. *Biological molecules in nanotechnology: the convergence of biotechnology, polymer chemistry and materials science*, edited by S.C. Lee and L. Savage (IBC Press, Southborough, MA).

Matsuki, H., K. Ofuji, N. Chubachi, and S. Nitta. 1996. Signal transmission for implantable medical devices using figure-of-eight coils. *IEEE Trans. on Magnetics.*, vol. 32. No.5, 5121. September.

Matsuki, H., Y. Yamakata, N. Chubachi, S. Nitta, and H. Hashimoto. 1996. Transcutaneous DC-DC converter for totally implantable artificial heart using synchronous rectifier. *IEEE Trans. On Magnetics.* 32, No.5, 5118 (Sept.).

Merz, J.L. 1986. "The Optoelectronics Joint Research Laboratory: Light Shed on Cooperative Research in Japan." *Scientific Bulletin* (ONR Far East Office). 11 (4), 1-30 (Oct.-Dec.).

NSET. 2000. *National Nanotechnology Initiative: Leading to the Next Industrial Revolution.* 2000. A report by the Interagency Working Group on Nanoscience, Engineering and Technology (NSET), of the National Science and Technology Council's Committee on Technology. Washington, D.C., February.

Pollock, C.R. 1995. *Fundamentals of Optoelectronics.* Irwin, Inc. pp. 3-5.

Porod, W., C.S. Lent, G.H. Bernstein, A.O. Orlov, I. Amlani, G.L. Snider, and J.L. Merz. 1999. Quantum-dot cellular automata: computing with coupled quantum dots. *Int. Journal of Electronics* 86, No.5, 549-590.

Roco, M.C., R.S. Williams, and P. Alivisatos (eds). 1999. *Nanotechnology Research Directions: IWGN Workshop Report, Vision for Nanotechnology R&D in the Next Decade.* (Kluwer Academic Publishers. Also available at http://www.nano.gov/).

Smalley, R.E. 2000. Hearing of the Senate Science and Technology Caucus on Nanotechnology. April 5, (unpublished).

Von Arx, J.A. and K. Najafi. 1997. On-Chip Coils with Integrated Cores for Remote Inductive Powering of Integrated Microsystems. *Proceedings of Transducers '97*, an International Conf. On Solid-State Sensors and Actuators, Chicago, June (published by the IEEE). p. 999.

DYNAMICS OF THE EMERGING FIELD OF NANOSCIENCE

H. Glimell, Göteborg University

Introduction

To follow is a selection of aspects considered important to "the dynamics" of nanoscience and nanoengineering (NSE). They reflect my picture of what counts in the community of NSE, while addressing approaches which social science could adopt to take account of and stimulate reflections on the "multi-faceted technoscience endeavor" preoccupying that community. Although I myself set out by organizing my aspects into three categories — cognitive, social and culture dynamics — the implied research agenda to emerge defies rather than advocates such a classification. In the concluding paragraphs, it is suggested that perhaps only by playing down another resistant demarcation in our society, namely the one separating professionally authorized expertise from lay expertise, will it become possible to envision and exploit the entire dynamics of nanoscience.

Cognitive Dynamics

Obviously, the expansion of nanoscience and the very idea of an emergent nanotechnology emphasize the need for interdisciplinarity. In spite of the positive response that challenge often invokes, the practice tends to be different. There is a long heritage within academia of the single discipline as the core entity of organization. Even in areas where external forces have exerted strong pressure to transcend them, disciplines have proven amazingly persistent. Also, whereas interdisciplinary undertakings often look like a win-win game from the outside, they typically on the inside are apprehended as a zero-sum game.

Within the field of science and technology studies (STS), Thomas Gieryn in the mid '80s introduced the notion of "boundary-work". Scientists sustain the epistemic and cultural authority of science by constantly drawing and redrawing maps or spatial segregations highlighting contrasts to other kinds of knowledge, fact-making methods, and expertise. Since then many forms of boundary-work (expulsion, expansion, protection of autonomy) have been shown to apply also to how credibility and cultural authority within science is being contested. When future cartographers try to come to grips with the many negotiations and redrawing of maps that we could expect to be a salient feature of NSE in the years to come, this is a vein of the social study of science to be consulted.

When research foundations launch a new "brand" on the R&D funding market — like now, when "nano" has entered the front stage through NNI — they also fuel a lot of new boundary work. The rules of the game are changed, and those who want to play it had better start depicting the new landscape (borders, barriers, shortcuts) and the emerging criteria which may give them access to it. At least two risks deserve our attention here. One is that a lot of resources go into the re-presentation and re-labeling of ongoing work, which nevertheless remains very much the same (it may also leave researchers with the bad feeling of having been forced to make promises they cannot fulfil). The second risk is that one succeeds in pulling new research into the field, but fails in securing for those new elements a sufficiently high quality. In Sweden, as an example, the dynamics of a major national energy research initiative launched in the early '80s were soon severely hampered by the disrepute of its quality deficiencies.

Several historians of science have brought our attention to the "sequential pattern" of the evolution of a scientific field. For example, from his thorough studies in the history of physics, Galison concludes that when there has been a breakthrough in one of the three innovation arenas he distinguishes — i.e. theory, (methods of) experimentation and instrument making — the other two stay stable until the new element has become fully assimilated in the scientific practices.

If such results are taken seriously, there should be implications here not only for historians but also for future-oriented policy makers — e.g., when they take on an emergent field such as NSE. Although of course aware that the above model is, for analytical reasons, a simplification of a much more interwoven reality, it can still draw the attention to a choice they have. Hence, they could choose a liberal or laissez-faire type of strategy; largely leaving to the research community to "make a bet" on where the next incremental change will occur. Or they may choose an interventionist strategy where funding is steered towards the arena they believe will generate the next major innovation impetus. Looked at from the individual research group's point of view, the strategic choice becomes one of either trying to encompass all three arenas or competencies, running the risk of allocating under-critical resources (i.e., no excellence) for everybody, or go for excellence in merely one of them, at the risk of not having picked the winner.

In addition to laboratory studies (or ethnographies of the everyday practice of science) and historical case studies, there is within STS a field that elaborates controversy analysis as a fine-tuned methodology with which the micropolitics for winning cognitive or epistemological credibility and hegemony is studied. As NSE in many respects represents a juvenile area, it is likely to see the coming-about of several scientific controversies. Some of these, if carefully studied, could become rich sources of knowledge on how the main demarcations and topography (cf., "boundary work" above) of the nano landscape is evolving.

Social Dynamics

Controversy studies do not only shed light on the internal consistency or credibility of the scientific claims under study. Also, they point at how those claims are embedded in, or proactively "mobilize," various social and cultural factors. By basically broadening the symmetrical analyses used on controversies, there is by now an extensive body of research on "the anatomy" of the wider societal networks constituting and channeling innovation. For example, one finds here the studies informed by the actor-network theory (ANT, or the "Paris school" founded by Bruno Latour), in which a "follow-the-actor" methodology guides the analyst and where "science" or "technology" take the shape of heterogeneous networks blending humans and non-humans, facts and fiction, the "technical" and the "social," "nature" and "culture."

Given the generic character of NSE, the close monitoring that ANT and similar approaches set in motion are useful in articulating the diversity underneath that character. Consider for example molecular electronics compared with bio-nano (or the interface of biological and organic nano materials). The actors, nodes and connections to appear in the extension of these NSE subareas obviously constitute two very different networks of innovation. Nanoelectronics is being negotiated and molded in between two camps — the conservative mainstream of the microelectronics industry with its skepticism towards anything popping up as a challenger to the three decade old CMOS technology trajectory, and the camp committed to a scenario where that trajectory might come to its end within some five years from now. As different and even antagonistic as those two camps may be, they are still very close from a cognitive point of view — i.e., they are perfectly aware of what the other is up to and why.

The bio-nano case is not like that. The corresponding dual relationship is here the one between the bio-nano scientists with their "wild ideas" of new hybrid materials to be applied in bio-interfaces and bio-implants, and the mundane health care sector with its clinics. There, the practitioners usually have great difficulties in grasping what the other party is so concerned about. A gap in terms of knowledge and everyday experience, rather than one of different assessments of the technology, here separates the two.

It is by no means clear which one of these two makes the better partner in bringing radical science towards applied innovations. What should be clear, however, is that only with a thorough knowledge of the often very different actor-actant-networks in action in various regions of the nano territory, the chances of "tailoring" initiatives, interventions and resources to the crucial nodes or points of communication and social interaction will improve.

A very crucial point of exchange, where the different character of nano networks matters a great deal, is where "government money" should be shifted over to "market money". Obviously these two imply very different expectations, rationalities and infrastructures. But interestingly, in some regions of the "nanoscape," science with government money is doing what industry normally should be doing; while in other regions, industry with its market money is doing what science should deliver. Within NSE, being in a premature state, many such "imperfections" will occur. Economists have studied some of these (e.g. in venture capital studies), but the tools they offer are at the same time too narrow and too crude to account for the complexity, importance, and social dynamics of these points of exchange.

Cultural Dynamics

The nano endeavor is in profound ways culturally embedded and relevant. It is spirited by the grandiose idea of man becoming as brilliant as nature in building things (materials, devices, artifacts, products), a truly utopian idea within a longstanding tradition of man's fascination over the prospects of science and technology. The "nano" is a lot of laborious work, but also it is nothing less than the ultimate challenge; the dream of any dedicated engineer. No surprise then, it is well established in popular culture. Long before it reached Presidential committees or NSF, it flourished in science fiction movies, in the fantasy literature, and in groups of radical "nanoites" organizing themselves to meet in VL or RL to discuss nano.

For at least a decade, the scientific community did boundary work to keep the demarcations towards this radical popular version of nano straight. In lacking some of the formal merits that make up the credentials of this community, and, even worse, in refusing to keep the distance from the non-scientist nanoites, Eric Drexler of course had to remain unauthorized "on the other side." By "demarcating" like this, one went free from both the worst "technophilia" and the "technohorror" of the nano discourse. In doing so, however, one missed the opportunity to conquer determinism by suggesting more modest or reflected understandings — in exactly the way demonstrated by John Seely Brown and Paul Duguid when they at this workshop take issue with the "tunnel vision thinking" guiding Bill Joy in writing his widely recognized *Wired* article. (By the way, illustrating the above demarcations, ten years before Joy's article, quite a similar analysis was presented without much notice at a Foresight Institute conference.)

One of my informants, a nano-bio scientist, recently told me how he believes that an extended interaction with people from the social sciences and humanities at early stages of the research process would make it possible for us to "lie ahead of ourselves." He belongs to those (still a clear minority I reckon) who would give full support to a meeting like this.

Fine; but every cultural analysis deserving its name should encompass reflexivity. Thus, coming to think about it: what does it really mean that we — a mixture a people from industry, government and universities — gather here for a few days to discuss "nano" before we know what actually will become of it? This is an important question, as we from the very moment we start discussing (and even from when we started to prepare ourselves by writing statements like this one) can be pretty sure of (without therefore boosting ourselves) affecting the development of "nano." How? We don't know, and we cannot know. The answers to most of the questions to be raised during our workshop can only be answered by future historians of science and technology. We will affect things, but we are not able ourselves to recognize our impact even when we see it. As once expressed by Jorge Luis Borges: "Time is a river, that pulls me along, but I am the river."

Beyond Business as Usual

Although I with everybody else share the predicament captured by Borges, my concluding words will still have an activist flavor. I do think that one of the most thrilling things about this workshop is that we well might be involved in something historical here. Ten or twenty years ago an event like this didn't happen. There was certainly a debate on the societal and ethical implications of technology in general, and there were foresight or scenario activities taking place, but the idea to actively try to integrate at a very early stage of a new emerging technoscience area also the perspectives and experiences of social scientists, is to my knowledge a new one.

Herein lies perhaps the real challenge and dynamics of the nano initiative. It could well be that NSE develops into something of a "public expertise field," where the vital research ahead of us becomes the concern of "the many," without of course therefore lessening the importance of science in the traditional sense. The vivid debate on the genetic technologies during the last few years has shown that "the public" may be prepared, and often increasingly well prepared, to take a greater responsibility for the science and technology underway (taking what is often referred to as "the public understanding of science" a step further). We don't know; perhaps the molecular biology revolution will come out too strong to give room for "nano" as another major issue in the debate. But also, this very much will depend on how the spokesmen for and members of the growing NSE community will respond to the possibility of planting "their" science and technology in the wider realms of the public.

Reference

Conversations with Professor Magnus Willander, Physical Electronics and Photonics; and Professor Bengt Kasemo, Chemical Physics; the Department of Physics, Chalmers Institute of Technology.

6.4 FOCUS ON MEDICAL, ENVIRONMENTAL, SPACE EXPLORATION AND NATIONAL SECURITY IMPLICATIONS

CHALLENGES AND VISION FOR NANOSCIENCE AND NANOTECHNOLOGY IN MEDICINE: CANCER AS A MODEL

R.D. Klausner, National Cancer Institute

The recognition of the molecular basis of cancers creates the opportunity for a future where cancers are detected, diagnosed, and treated based on the fundamental changes in the specific disease. Ongoing efforts target the definition of the genes and expressed gene products of the human genome, the discovery of sentinel biomarkers of the early presence of disease, and establishment of informative diagnostic classification systems based on the fundamental molecular changes. Closely linked are efforts to discover and exploit molecular targets for cancer prevention and treatment. New technologies are needed to speed the discovery process that are rapid, highly parallel, and cost-effective. Reductions in the scale of analysis tools and automation are proving critical to enabling the discovery of the fundamental changes associated with the development of cancers, and insight into the molecular processes of the cell.

The identification of molecular signatures of cancers will enhance our ability to identify and accurately diagnose disease. Our goal is to use this information to identify cancers or precancers at the earliest point in the disease process and intervene before symptomatic disease becomes apparent. Realization of this goal will only be possible if technologies exist that allow us to scan the living body for the earliest signatures of emerging disease and support immediate, specific intervention. The ability to scan the body for early signatures requires these technologies to be minimally invasive. The ability to scan the body for molecular signatures will also allow us to monitor the progress of disease and effects of interventions.

The NCI is currently seeking technologies that will support the earliest detection of the molecular signatures of cancer and serve as a platform for the seamless interface between detection, diagnosis, and intervention. Platform technologies must integrate the ability to sense, signal, respond, and monitor. Nanotechnologies are emerging as enabling components to these goals. Ongoing efforts highlight that full systems will require the integration of new discoveries from a variety of fields including nanoscience, chemistry, photonics, computational sciences, and information science and technology.

NANOTECHNOLOGY IN MEDICINE

S.I. Stupp, Northwestern University

Nanotechnology, which can be presently viewed as the promise of a technology, could have a profound impact in medicine in the not too distant future. But incorporating now nanoscience in the national research agenda for advanced medicine is important not only for the obvious reason but also because this enormous challenge will impact science broadly, and help us make the needed cultural transition to interdisciplinarity. Nanotechnology in medicine is about the hybridization of physical sciences, biology, engineering, and clinical medicine. It is difficult to identify an area that requires this much synergistic mixing of traditional fields. This fact poses an educational challenge at all levels.

The promise of nanoscience for medicine rests on various grand challenges. An important one is connected to our abilities to manipulate the behavior of a "single cell" or groups of cells of common phenotype using synthetic nano-objects that are targeted to interact specifically with the cell's own functional nano-objects (i.e., receptors, cytoskeleton parts, specific organelle locations, nuclear compartments, etc.). Into the future this area will allow us to diagnose disease at much earlier stages than we do presently, reverse disease, repair or re-grow human tissues, maybe enhance human performance when needed (this of course touches on complex dilemmas for society not commonly addressed by scientists).

Among other things, the challenge is related to what chemistry does not yet do well at all, teach the synthesis of well-defined objects in the 1-100 nanometer range. This will require also understanding and inventing new modes of molecular recognition, new tools to manipulate and detect the presence of exceedingly small numbers of nano-objects such as proteins inside and outside of cells. Engineers and physicists will play key roles in the development of these tools, but of course the users, clinicians, will have to guide and help prototype nanotechnologies in medicine. Finally, it is clear that advances in nanoscience for medicine will impact in parallel other fields such as environmental detection of toxins and pathogens and maybe our abilities to manipulate agriculture at a level not yet experienced.

LIFECYCLE/SUSTAINABILITY IMPLICATIONS OF NANOTECHNOLOGY

L.B. Lave, Carnegie Mellon University

Abstract

Nanotechnology has the potential to be the new technological wave that increases environmental quality and sustainability. By reducing the amount of energy and materials required to accomplish a desired task, nanotechnologies can provide the goods and services we desire "smarter, cheaper, faster" and with a smaller environmental footprint. Transistors and later microprocessors are part of the evolution of accomplishing our consumption goals with less energy and materials. A standard graph shows the rapid decline in the cost of memory or logic. The graph might be reconfigured to show the decline in the amount of energy, materials, or environmental discharges required to perform a calculation or store a byte. Nanotechnology may be the next step. It promises to reduce by orders of magnitudes the inputs of energy and materials and associated environmental discharges required to produce a device that can perform a particular task. The result could be perhaps an order of magnitude increase in real income for the current world population without requiring more energy, materials, or resulting in additional discharges. Thus, nanotechnology offers the prospect of giving poor nations much higher standards of living and making the world economy sustainable.

In biological systems, the population of a species will continue to grow as long as there is food and other conditions that permit growth. No individual, or the species, evidences concern that the population might be growing too large to be sustainable. Often, the expansion results in over-population and a crash with many individuals starving.

A widely read book of the mid 1970s, *Limits to Growth*, assumed the same behavior would occur for human populations. It assumed that the population and use of energy and materials would continue to grow exponentially. Meadows et al. (1972) found that, eventually, there would be a catastrophic crash, as fossil fuels and raw materials were exhausted. The book demonstrated the truth that exponential growth is inconsistent with a finite world.

Two centuries ago, Thomas Malthus explored the same notions, concluding that humans would continue to breed until starvation, pestilence, and war limited the population size.

Income in England is far greater than in Malthus' time. While there is starvation, this seems to have more to do with conflict than any inherent shortage of food (Simon 1995). Life expectancy has increased in all the developed nations and in almost every part of the world. Thus, aside from human conflict and a relatively small burden of disease and natural disasters, people live much better than Malthus predicted.

Malthus and Meadows et al. (1972) erred by not accounting for the effects of technological change, with its inherent ability to substitute abundant materials for scarce ones. They also erred by not accounting for the feedback in a market economy. Scarcity causes increasing prices, which signal inventors to find substitutes, prospectors to find other supplies, and consumers to use less or find substitutes.

Humans are different from other animals in explicitly and implicitly accounting for the systems wide impacts of their actions, including having more children. Even when consumers don't think about the economy-wide effects of increased population, a market will react to increasing population by increasing the price of scarce food and other materials and by lowering family income (by decreasing the wage rate, or increasing unemployment). Both effects lower the standard of living and cause people, eventually, to think about the desirability of having so many children. The feedback effect can take decades and the signals often are misinterpreted, but the market signals get stronger as population outpaces the ability to support the population. Thus, humans are fundamentally different from bacteria, plants, and other animals. We have institutional feedback systems that prompt people to reassess their decisions to have more children before there is a disaster.

The importance of technological change cannot be overstated. Humans started out as foragers, dependent on the bounty of nature. To support a growing population and to ensure a steady food supply, humans developed agriculture. Agricultural techniques developed over time, including breeding more productive plants and animals, irrigation, mechanization, and the use of fertilizers and pesticides.

Wood was a primary energy source for cooking and smelting metals. As the forests of Europe were depleted, people hunted for, and eventually found a substitute in coal, and later in petroleum and natural gas. Mining coal, transporting it, using it to reduce iron ore, and learning to burn it without excessive pollution required vast improvements in technology. Finding and producing petroleum and natural gas posed still greater challenges to technology.

Rich deposits of ores for tin, copper, lead, and iron were gradually exhausted in Europe (Tilton 1991). Technology developed to produce the metals from less rich ores and to find substitutes. Iron ore is more plentiful than ores for tin, lead, and copper (used to make bronze). Tin cans for storing food compete with glass jars. The competition among materials led to recycling as well as thinning the layer of tin in order to lower cost. Technology was called upon to use expensive resources more productively and to find substitutes. Similarly, the scarcity of natural rubber during World War I led to the development of synthetic rubber, made from petroleum. Petroleum shortages in Germany during this war led to the development of synthetic fuels made from coal.

How Many People Can the Earth Support?

People such as Ehrlich (1977) have been deeply concerned about over-population and the ability of the Earth to feed increasingly large populations. The Earth currently supports 6 billion people. As noted above, aside from political problems, wars, and occasional natural disasters, the 6 billion people have enough to eat and many are far above subsistence level. This is not saying that the quality of life is high for all 6 billion people, since there would be less crowding with fewer people.

It seems self-evident that the Earth could not support 6 billion people if they were all foragers. There are not enough roots and berries and game to support so many people. If foraging were the technology for feeding people, most would starve.

Similarly, if the technology were early agriculture with its primitive grains, lack of pesticides, fertilizers, and irrigation, most people would starve. There simply is not enough good farmland to support so many people without the improvements that have come from plant breeding, pesticides, fertilizer, and irrigation.

The standard of living in the United States is far above subsistence because only about 2% of the workforce is required to produce the food. Particularly in the last century, the combination of technological change and clever use of resources has outpaced population growth, especially in the developed nations. However, current agriculture, and the economy more generally is built on a foundation of fossil fuels, underground aquifers, and other resources that are very finite. If current technology were used to provide all 6 billion people with American lifestyles, we would quickly exhaust petroleum resources, ores, water supplies, and pollute the environment. Can technological change and clever resource use continue to outpace population growth? Will world population growth slow and then be transformed into decreasing world population? I am optimistic, but the world of 2100 is unknowable.

Nanotechnology Opportunities

Burning fossil fuels is the primary source of pollution, including polluted air and greenhouse gas emissions in the USA. Nanotechnology offers the promise of being the foundation for the next wave of technological change. The energy from fossil fuels is used extremely inefficiently. Nanotechnology could provide improved services for a small fraction of current energy in lighting, computing, printing, water filtration, and many other areas. For example, only a few percent of the energy in gasoline is actually used to get me where I want to go. The current internal combustion engine is about 15% efficient in transforming the energy in the gasoline into power to turn the tires. In addition, the average car weights about 3,000 pounds and is transporting one 160-pound person. There is no reason in principle why converting the energy in gasoline into useful work could not be five times more efficient. Similarly, there is no reason in principle why the amount of non-useful material transported with me could not be reduced by a factor of ten. If so, person transportation vehicles could be 50 times more efficient than at present.

The Dark Side of Nanotechnology

Every intervention in natural systems has undesired consequences. For example, the historian Lynn White (1974) asserts that the cause of the French Revolution was the invention of chimneys. Prior to chimneys, every member of the household slept around a fire in a room with a hole in the roof. The invention of chimneys permitted individual rooms to be heated and vented, allowing the rich to distance themselves from their servants and the poor, eventually leading to the excesses of Louis XIV and the French Revolution.

In a far more direct way, new technology has undesired consequences that can nullify its advantages or at least require considerable changes. For example, asbestos is a marvelous insulator and wonderful at fireproofing. However, the small fibers cause asbestosis, lung cancer, and mesothelioma. Despite its wonderful properties, asbestos has been banned in the USA.

To assess a nanotechnology, a lifecycle analysis is needed. Are there hazardous materials produced? Do parts of the product cause safety hazards? Are the needed materials in abundant supply?

The Green Design Initiative has evaluated some technologies that are well thought of, or even required by regulators. The conclusions can be surprising. For example, the California Air Resources Board initially required that 2% of new cars sold in 1998 had to be zero emissions vehicles. The only vehicles that satisfy the requirement are battery-powered vehicles. Lave et al. (1995b, 1996), found that mining and smelting the lead and then making and recycling the 500 kg battery in an EV-1 would result in 4-60 times as much lead being discharged into the environment, per vehicle mile, as a comparable car using leaded gasoline. Similarly, large amounts of nickel and other heavy metals would be discharged into the environment if the batteries were made from nickel metal hydride. Another evaluation found that, while hybrid-electric cars are more fuel efficient and less polluting than cars using a conventional internal combustion engine, the differences are small and not commensurate with the increased production cost (Lave and MacLean, 2000). While it advances technology, the Toyota Prius hybrid electric vehicle is not cost-effective.

The Potential for Nanotechnology

One idea of the potential for nanotechnology can be seen in Table 6.1. This table uses our lifecycle software (Lave et al. 1995a; Hendrickson et al. 1998) to compute the resource use and environmental discharges of several sectors. (This software was developed, in part, with NSF funding and is available free on the web at www.eiolca.net.)

The table lists the amount of materials, energy and environmental discharges associated with purchasing $1 million worth of that output. The first column lists the categories that we tabulate for each sector. These include total energy, various fuels, ores, water, and various categories of environmental discharges. The second column is the electricity generation sector of the United States. This sector uses a great deal of coal, natural gas, petroleum, nuclear and hydropower, and small amounts of renewable technologies, such as wind power. Since it uses such large amounts of fossil fuels, it generated a great deal of environmental discharges, particularly into the air.

The total amount of energy used to generate $1 million of electricity is 147 terajoules. The electricity sector purchases 0.1-megawatt hour of electricity. It releases 105 metric tons of conventional air pollutants and 12,570 metric tons of greenhouse gases (carbon dioxide equivalent). Producing the electricity results in 7/10,000 of a fatality. Generation uses 4,997 metric tons of fuel and 28 metric tons of ores (to produce the steel, etc. for the plant). Generation results in 15 metric tons of RCRA hazardous waste and 2 tons of toxic releases and transfers. The external cost of releasing the air pollutants is 33.9% of the total revenue or $339,000. Finally, generation uses 1 million gallons of water.

Table 6.1. Life Cycle Implications of $1 million Sale by Sector

Effects	Electricity	Steel	Aluminum	Computers	Computer Services
Electricity Used [Mkw-hr] x 10	1	19	138	3	1
Energy Used [TJ]	147	74	54	7	3
Conventional Pollutants Released [metric tons]	105	52	106	5	3
OSHA Safety [fatalities]x10,000	7	9	7	4	3
Greenhouse Gases Released [metric tons CO_2 equivalents]	12570	6140	3002	460	226
Fuels Used [metric tons]	4997	2443	1165	175	85
Ores Used [metric tons]	28	370	323	172	35
Hazardous Waste Generated [RCRA, metric tons]	15	75	76	32	10
External Costs Incurred: as Percentage of Revenue	34	14	15	1	1
Toxic Releases and Transfers [metric tons]x10	2	72	27	9	2
Water Used [million gallons]	1	66	18	2	1

Clearly, conserving electricity would be worth a great deal toward sustainability and environmental quality.

The next two columns show the implications of purchasing $1 million of steel or aluminum. Aluminum uses much more electricity than steel, but steel uses more total energy (twice as many tons) and so releases more greenhouse gases. The sectors are similar in terms of ores used, hazardous wastes, and external costs. However, steel results in more toxic releases and transfers and water use. Because it uses so much electricity, aluminum is responsible for twice the emissions of air pollutants. The industries are comparable in terms of work related fatalities.

Column 5 shows the implications of buying $1 million worth of electronic computers. This sector is much more benign than the previous three sectors.

The final column shows the implications of buying $1 million of computer and electronic data processing services. This sector is still more benign than producing computers, since it performs a service rather than making a product.

For the computer sector, about 20% of the energy and materials use and of the environmental discharges are due to the sector directly, while all the rest is due to industry suppliers. For computer services, only about 4% is due to the sector itself, with 96% due to suppliers. Thus, if both sectors could use less electricity and steel and aluminum, their environmental footprint would be appropriately smaller. In particular, if nanotechnology enabled electricity to be generated with less fuel and less environmental discharges, that would make a huge contribution to decreasing the environmental footprint of this sector. Similarly, if computers could be made with less energy and raw materials, it and computer services would be enormously benefited.

These lifecycle calculations demonstrate the potential benefits for improving environmental quality and sustainability. Nanotechnology could make a large contribution to lower resource use and environmental discharges.

Comment in Plenary Discussion

The general public has become increasingly concerned about the safety and environmental implications of new technologies, from nuclear power to chemicals to biotechnology. Statutes such as the Toxic Substances Control Act and Food, Drug, and Cosmetic Act require government review of new chemicals and other technologies before they can be sold. Although proponents of nanotechnology view it as benign, there are likely to be some unforeseen, undesirable effects.

Even at the basic research stage, nanotechnology advocates need to inform the public about the prospects and risks. They need to engage and involve the public and the groups that represent them. While this will delay the introduction of new technologies, in the end it is likely to save time.

To win support for this initiative from Congress and the general public, nanotechnology advocates need to specify the social problems that can be addressed by nanotechnology. They need to make a convincing case that expenditures will be more productive in addressing these problems than would be expenditures on a variety of other social programs, from paying down the debt to tax cuts to Medicare prescription benefits.

In communicating with Congress and the general public, it is important to use their language and to make the communication understandable.

The social sciences have much to offer in addressing issues of evaluation, communication, and addressing social needs.

References

Ehrlich, Paul R., Anne H. Ehrlich, and John P. Holdren. 1977. *Ecoscience: Population, Resources, Environment*. San Francisco: W.H. Freeman.

Hendrickson, C.T., A. Horvath, S. Joshi, and L.B. Lave. 1998. Economic input-output models for environmental life-cycle assessment. *Environmental Science & Technology*. p. 32.

Lave, L.B. & H.L. MacLean. 2000. *An Environmental-Economic Evaluation of Hybrid Electric Vehicles: Toyota's Prius vs. Its Conventional Internal Combustion Engine Corolla*. working paper. August.

Lave, L.B., A. Russell, F.C. McMichael, and C.T. Hendrickson. 1996. Environmental implications of battery powered vehicles as a transitional technology: ozone vs. lead. *Environmental Science & Technology*. p. 30.

Lave, L.B., C.T. Hendrickson, and F. C. McMichael. 1995b. Environmental implications of electric cars. *Science*. vol. 268: 992-5. May 19.

Lave, L.B., E. Cobas-Flores, C.T. Hendrickson, and F.C. McMichael. 1995a. Using input-output analysis to estimate economy wide discharges. *Environmental Science & Technology*, 29(9).

Meadows, D.L., et al. 1972. *The Limits to Growth*, New York: University.

Simon, Julian L. (ed). 1995. *The State of Humanity*, Oxford: Blackwell.

Tilton, John E. (ed). 1991. *World Metal Demand*, Washington, DC: Resources for the Future.

White, Lynn Jr. 1974. Technology assessment from the stance of a medieval historian. *The American Historical Review*. 79(1): 1-13.

IMPLICATIONS OF NANOTECHNOLOGY FOR SPACE EXPLORATION

S.L. Venneri, National Aeronautics and Space Administration

Abstract

Typically when NASA begins a new technology program the Agency is most concerned about the performance benefits, cost of development, time for development and new opportunities that are enabled. Ethics becomes part of the process if the development or ultimate use of the technology directly affects the health or well being of humans or other living creatures. However, as we move into the era of nanotechnology we are also encompassing biology and fundamental biological processes. Our vision of nanotechnology encompasses the attributes of self-generation, reproduction, self-assembly, self-repair and natural adaptation. These are all attributes we ascribe to living things. Thus, we are moving beyond the typical bounds of technology into the domain of natural philosophy. This can have significant implications for the public attitude toward such technology.

Nanotechnology will enable NASA to build future systems with many of these "life-like" characteristics. We need this capability for our robotic systems to operate at great distances from Earth, in harsh environments without the benefit and high cost of continuous human control. As we develop new nanotechnology we must also pro-actively establish policies and guidelines to assure the technology and systems made from it are socially acceptable to the general public.

Ethics as a Decision Criteria for Technology Planning and Development

During the early days of the Space Age, the United States was forced to develop spacecraft that were small but powerful. We lacked the large launch vehicles that could put heavy payloads into orbit so we concentrated on developing miniature systems. The microelectronics revolution was a product of this era. However, by the 1970s the spacecraft NASA was building for deep space missions had grown to weigh thousands of kilograms. This trend continued into the early, 1990s. Viking, Galileo, Voyager, Magellan and Cassini all weighed thousands of kilograms at launch. These missions also cost billions of dollars. Though much of this weight was in propellant, the total spacecraft weight was driven by the size of the final payload.

In the early 1990s NASA moved away from large expensive missions and focused on developing spacecraft an order of magnitude smaller and less expensive. In addition, NASA increased its efforts to develop on-board "intelligence" to reduce the cost of operations. For some of NASA's missions, the cost of maintaining an "army" of operators to monitor and control every critical function of the spacecraft was comparable to the cost of the spacecraft itself.

This move toward smaller, smarter, lower cost systems has become essential if the Agency is to accomplish its future missions. As we look to the future, all of the "easy" missions have been accomplished. We have flown by every planet except Pluto and orbited Venus, Mars and Jupiter. Cassini will eventually orbit Saturn. The only planetary body we have explored in person is the moon — and even that for less than three weeks in total. Space exploration of the future will be characterized by more in-depth investigations. We will put spacecraft on the surface of planets, moons, asteroids and comets. We will explore below the surface and in the atmosphere as well. Closer to home we will explore the structure and dynamic behavior of the Earth's geomagnetic environment and distribute spacecraft in strategic locations to learn about the intimate relationship between the Earth and the sun.

One of our highest priorities is to search for life elsewhere in the solar system. While space exploration of the past may have excited the country, the future looks even more exciting. We have recent evidence of liquid water near the surface of Mars and there is the real possibility of more water below the icy surface of Europa than in all the oceans on Earth. And, on Earth everywhere we have found water we have also found life, so these recent discoveries are all the more exciting. While our initial "explorers" will be robots, if our solar system proves a sufficiently vibrant source of scientific discovery human explorers will eventually follow.

However, our current challenge is to develop space systems that can accomplish our missions effectively and economically. To do this they will have to be much more capable than today's spacecraft. They will have to have the characteristics of autonomy to "think for themselves"; self-reliance to identify, diagnose and correct internal problems and failures; self repair to overcome damage; adaptability to function and explore in new and unknown environments; and extreme efficiency to operate with very limited resources. These are typically characteristics of biological systems, but they will also be the characteristics of future space systems. A key to developing such spacecraft is nanotechnology.

We are already seeing the potential of nanotechnology through the extensive research into the production and use of carbon nanotubes, nano-phase materials and molecular electronics. For example, on the basis of computer simulations, and available experimental data, some specific forms of carbon nanotubes appear to possess extraordinary properties: Young's modulus over one Tera Pascal (five times that of steel) and tensile strength approaching 100 Giga Pascal (over 100 times the strength of steel). Recent NASA studies indicate that polymer composite materials made from carbon nanotubes could reduce the weight of launch vehicle — as well as aircraft — by half. Similarly, nanometer-scale carbon wires have 100,000 times better current carrying capacity than copper, which makes them particularly useful for performing functions in molecular electronic circuitry, now performed by semiconductor devices in electronic circuits. Electronic devices constructed from molecules (nanometer-scale wires) will be hundreds of times smaller than their semiconductor-based counterparts. However, the full potential of nanotechnology for the systems NASA needs is in its association with biology.

Nanotechnology will enable us to take the notion of "small but powerful" to its extreme limits. Biology will provide many of the paradigms and processes for doing so. Biology has inherent characteristics to enable us to build the systems we need: selectivity and sensitivity at a scale of a few atoms; the ability of single units to massively reproduce with near zero error rates; organization capability to self assemble into highly complex systems; the ability to adapt form and function to changing conditions; the ability to detect damage and self repair; and the ability to communicate among themselves. Biologically inspired sensors will be sensitive to a single photon. Data storage based on DNA will be a trillion times more dense than current media; and supercomputers computers modeled after the brain will use as little as a billionth the power of existing designs. Biological concepts and nanotechnology will enable us to create both the "brains and the body" of future systems with the characteristics we need. Together, nanotechnology, biology and information technology form a powerful and intimate scientific and technological triad.

An example is intelligent multifunctional material systems consisting of a number of layers, each used for a different purpose. The outer layer would be selected to be tough and durable to withstand the harsh space environment, with an embedded network of sensors, electrical carriers and actuators to measure temperature, pressure and radiation and trigger a response whenever needed. The network would be intelligent. It would automatically reconfigure itself to bypass damaged components and compensate for any loss of capability. The next layer could be an electrostrictive or piezoelectric membrane that works like muscle tissue with a network of nerves to stimulate the appropriate strands and provide power to them. The base layer might be made of bio-molecular material that senses penetrations and tears and flows into any gaps. It would trigger a reaction in the damaged layers and initiate a self-healing process.

Such systems will use the design and fabrication methods very different from those we use today. Today, we build most of our systems by starting with volumes of material and "chipping" away what we do not need, or by selectively layering material over "large areas" — large compared to the scale of the phenomenon we are trying to control. Doing so, we can produce several million transistors on a single microchip. But we are still limited by our ability to cut or to layer. By contrast, biology intrinsically works at the atomic level and builds systems far more complex than anything we can build today, atom by atom. Such nanoscale systems can be 10,000 times smaller than current systems.

This same technology will also enable us to send humans into space with greater degrees of safety. While the vehicle they travel in will have much greater capability — and display the same self-protective characteristics of spacecraft — nanotechnology will enable new types of human health monitoring systems and health care delivery systems. Nanoscale, biocompatible sensors can be distributed throughout the body to provide detailed information of the health of astronauts at the cellular level. They will have the ability to be queried by external monitoring systems or be self-stimulated to send a signal, most likely through a chemical messenger. NASA is currently working with National Cancer Institute to conduct research along these specific lines.

The societal implications will not just be new and exciting space exploration missions. As in the past, the demands of space exploration have resulted in scientific and technological advances with great benefit to the country in general. The communication satellites we depend on so heavily today are products of the country's space program. The monitoring systems used in intensive care units and in heart rehabilitation wards are descendants of the systems used to monitor the heart beat of astronauts during the first space missions in the early 60s. Today we take such technology for granted.

Recently scientists have found tiny biological motors naturally occurring within cells. Their biological function is to help the cell generate energy, but they are also amazing little machines. They look like a tiny ring of footballs with a broomstick in the middle. Each of the components is a large complex molecule. In the process of generating energy for the cell the "shaft" in the center of the "motor" spins, like a microscopic electric motor.

NASA is supporting research to employ these biomolecular motors as the power source for fabricated nanomechanical devices. These devices are fueled by the chemical sources that provide energy to the cells and thus can potentially be safely and seamlessly integrated with a living host. One of the potential uses for these devices is to create cellular pharmacies that could dispense medication directly into individual cells. They would be coupled to nanosensors that would detect when medication is needed and dispense exactly the right amount molecule by molecule. Because the biomolecular motors are fabricated using the machinery of life, there exists the very real possibility that we will be able to develop devices that are self-assembling and self-repairing.

The current era of nanotechnology is still in an embryonic stage and its full potential can only be speculated. But, combined with biology and information technology it can lead the way to a technical revolution as significant as Newton's laws of gravity and motion, electromagnetic theory of the 1800s or atomic theory, relativity and genetic discoveries of the past century.

However, a distinct difference is that as we merge nanotechnology with biology and information technology we will be building systems that become more and more "life-like" and which interact with and support living systems at the cellular level. On the positive side this will result in systems that more effectively meet our needs and communicate with us on our own level — for example, natural language. Sensory systems such as sight, sound and touch will mimic our own, though exceed human performance levels. This is what we envision for space systems.

But, there is a "down side" as well. As we proceed along this path we must be sensitive to the perception that our "life-like" technology and systems are actually "living" systems and that systems which are designed to interact with humans in a "human-like" manner may viewed as being "too human". In the past this has been the domain of science fiction; in the foreseeable future it could be reality. Our view at NASA is to be pro-active in developing ethical standards to make clear that we understand the accepted boundaries between true "life science" and "life-like" science. And, to make sure that our use of biological analogs and processes remains in the domain of technology. This will require us to engage others outside the fields of science and technology that NASA is most familiar with. It will not be sufficient to have approval of social and ethical leaders. We will have to maintain the acceptance of the general public. Currently, NASA has initiated a process to establish a task force under the NASA Advisory Council to specifically address ethics and technology.

NASA's role is explore the boundaries of aeronautics and space with machine and with people. As we do so the safety of our people is the highest priority, followed by the overall success of our missions. We will always use robotic systems wherever possible, and always when the safety of an astronaut cannot be adequately assured. Nanotechnology will provide the capability to build the small, compact systems that can perform tasks that today only humans can perform, and to do so economically and efficiently.

Nanotechnology holds great promise of revolutionizing space exploration with the effect of developing technology of great benefit to us on Earth. But, as we fully realize this potential we must proceed carefully.

NATIONAL SECURITY ASPECTS OF NANOTECHNOLOGY

W.M. Tolles, Naval Research Laboratory (Retired)

Abstract

Nanotechnology represents a wide spectrum of disciplinary and interdisciplinary research frontiers that will have a positive technological impact influencing our social and economic well being. National security, both economic and military, represents one vital aspect of governmental concerns that will be impacted by the anticipated discoveries and developments in this field. Competition for economic security represents an important aspect of nanotechnology while augmenting military security and defense capabilities. The development of nanotechnology offers much enhanced capabilities to the Department of Defense (DOD) in the performance of its mission. On the other hand, even though nanotechnology is at an embryonic stage in its development, "visionaries" have imagined powerful and, in some cases, frightening capabilities emerging from this technology. Many of these imagined capabilities are irrational, generated from hypotheses far removed from experiments and the laws of nature as we understand them today. Without any scientific bases, dire predictions of self-replicating species cause fear in the society that is already facing a threat from biological agents similar to that from other hypothetical self-replicating entities. Such fears are generated for a technology whose ultimate capabilities are not well understood at present. Many irrational fictional predictions represent a potential barrier to progress by raising imagined and very unlikely scenarios. A series of experiments and/or observations is proposed that should serve to alert the scientific community, and society in general, to take action when progress approaches the possibility that it poses a threat to society.

Introduction

National security involves the protection of our form of government and our way of life from internal and/or external threats, and involves maintaining stability in national and international functions such that violent force (or a threat of force) is not used to influence economic and social discourse. This involves not only open warfare, but includes localized conflict, terrorism, and a wide variety of combat scenarios as well as missions in peacekeeping. Maintaining a strong defense has been a key component of national security in the U.S. A keystone of U.S. defense posture includes maintaining a strong Science and Technology R&D program in order to have leading edge technologies available for timely weapons development as required. Nanotechnology represents one of these emerging technologies that can provide much needed enhanced capabilities to DOD.

The International Nature of Nanotechnology

To view nanotechnology in the proper perspective relative to national security, it is necessary to understand that research in nanotechnology is without national borders. As such, nanotechnology is an exciting research frontier pursued by many nations for more than a decade. Europe and Asia are strong competitors with the U.S. for advances in nanotechnology. As one way to gauge the level of efforts in other countries, the number of recent papers mentioning the word "nano" in their title in five regions of the world is given in Table 6.2. Although different nations may use the phrase "nano" in different contexts, with some deference to wording introduced by translation, it seems clear that the U.S. is second to Europe in activity involving nanotechnology. Further, Asia has a substantial effort rivaling that in Europe. The U.S. must compete with other nations in this hotly contested field of nanotechnology.

To a significant extent, economic security impacts on the security of a nation. Rapid declines in industrial competitiveness can lead to large levels of unemployment, unrest, and a basis for security issues. It is vitally important to maintain a competitive edge involving enterprises considered important for international competitiveness. The pervasive nature of nanotechnology research, and the important anticipated products that will influence future industrial products, implies the need for vigorous research programs to pursue the opportunities offered in this emerging technology.

Table 6.2. Number of Open Literature Articles, 1999-July 24, 2000[6]

Country	No. Articles on Nano	No. Articles In Database
China	1,120	41,175
Europe	3,309	550,427
Japan	1,121	124,195
Russia	472	44,644
USA	2,321	170,367

Contributions to National Defense

Nanotechnology is an appropriate term encompassing many research disciplines actively pursuing scientific frontiers involving matter at nanometer dimensions. The development of proximal probes (including the scanning tunneling microscope, atomic force microscope, and many derivative instruments) and other tools to examine the properties of matter at these dimensions have brought together several disciplines anxious to use these tools to advance the more traditional pursuits. Understanding and manipulation of materials at the atomic and nanometer scale has reached unprecedented capabilities in these disciplines. Contributions of nanotechnology to traditional defense systems will be many. It takes little imagination to elucidate developments that will lead to advanced materials, sensing and signal processing, information technology, battle management, casualty care, or medical procedures and medicines. These science and engineering fields will advance through the use of tools and new knowledge uncovered by research in nanotechnology. Additional benefits in other fields will also occur. Many excellent discussions involving the prospects of nanotechnology need not be repeated here (Roco 2000).

Advances in nanotechnology in the civilian sector will provide advantages to national security and military capabilities through "commercial off-the-

---

[6] Articles available in the Science Citation Index database were searched for these statistics. A more detailed examination was made involving 200 papers from each of five regions: China, Europe, Japan, Russia and the USA. The most recent papers from these regions that were documented by Science Citation Index during 1999-July 24, 2000 were chosen to obtain a statistical sample. Titles were examined that contained the word "nano-;" in almost all cases, about 90%of these titles represented efforts in nanotechnology as we consider it in the U.S. (excluding efforts with the word "nanoseconds," for example).

shelf" systems ("COTS") as well as through technologies and systems developed by defense laboratories. In many cases economic and military opportunities are considered to be complementary. This is a reflection of the basic theme by Paul Kennedy in his book, *The Rise and Fall of the Great Powers* (Kennedy 1987). After tracing the history of many nations, Kennedy reached the conclusion that a major consideration affecting the sustainability of nations was the need to keep national economic and military levels of effort in balance. This lesson must be remembered and remain a guiding light for the influence of nanotechnology on national security.

The traditional opportunities anticipated from research in nanotechnology (Roco 2000) may be rearranged in an alternative taxonomy to view the subject from a military perspective. At the present stage of discovery and development, these opportunities represent evolutionary improvements in military technology rather than any dramatically different approach. Many revolutionary capabilities are yet to be uncovered. Certainly, as we develop and implement many of the anticipated enhancements to traditional military defense, we must be aware that any potential adversaries have an equal opportunity for introducing such advantages. The competitive process involving technological superiority as it is applied to warfare has continued for centuries.

A preliminary list of these opportunities using terms and objectives more in line with a DOD posture includes the following:

1. Higher performance platforms (aircraft, ships, subs, boats and satellites) through stronger, lighter weight structural materials, stealth materials, and low maintenance and "smart" materials.

2. Enhanced sensing through more sensitive and selective sensors of electromagnetic radiation, magnetic and electric fields, nuclear radiation, and chemical/biological agents. Miniature systems capable of mobility and highly sensitive/selective sensors, combined with wireless communication, are envisioned for remotely determining the state of a potential battlefield.

3. Enhanced human performance through improved monitoring devices, even through an introduction of appropriate biological materials to enhance performance. Devices of all kinds to sense the state of a war fighter's physiological condition will enhance his/her effectiveness.

4. Information dominance through enhanced information technology. This is likely to take the form of smaller, lower power memories, smaller and

faster logic devices through improved processing, and enhanced secure communication systems with greater bandwidth.

5. Safer operation involving hazardous materials or operations, through the use of remotely operated robots.

6. Reduced manpower requirements through the greater use of automation in the maintenance, management and control of weapon platforms, systems, and hazardous functions.

7. Improved battlefield casualty care through the use of materials and procedures; for example: artificial blood substitutes, burn treatments, and biocompatible materials.

8. Battlefield remediation of chemically or biologically contaminated areas and/or equipment through the use of enhanced chemicals and procedures.

9. Lower life-cycle costs through the use of improved materials, coatings, and condition-based maintenance.

An active nanotechnology program involving researchers in the Department of Defense, spanning the activities in academia, industry, and defense, is necessary. It is by such programs that the necessary scientific knowledge and understanding is gained and transferred in an optimum manner. This breadth of R&D activity provides an efficient mechanism for military applications.

The Emergence of Irrational Visions

The question involving developments in nanotechnology, and whether they (along with sister research areas of genetics and robotics) represent destabilizing national security has become of increasing interest this year (Joy 2000). Visions of "smart" self-replicating miniature robotic systems easily manufactured by a rogue state (or terrorist group) have received considerable attention. "Visionaries" who have never performed experiments have nevertheless constructed scenarios raising highly questionable possibilities. Due to a lack of contact with reality, they envision a world in which ideal "machines" assemble atomically perfect systems having surprisingly "smart" capabilities. These systems are ostensibly not only capable of reproducing themselves, but are intelligent, and may be constructed to cause harm to the environment or living species. There seems little doubt that such systems are figments of imagination by very creative minds, and are nearly impossible based on the laws of physics, thermodynamics, or other laws of nature, as we understand them. Predictions of such occurrences, however, have caused so much concern that addressing the subject rationally is appropriate.

Many statements have appeared representing bizarre predictions resulting from uncontrolled progress in nanotechnology. The following is one such example:

> The possible applications of nanotechnology to advanced
> weaponry are fertile ground for fantasy. It is obvious that three-
> dimensional assembly of nanostructures in bulk can yield much
> better versions of most conventional (non-nuclear) weapons; e.g.,
> guns can be lighter, carry more ammunition, fire self-guided
> bullets, incorporate multispectral gun sights or even fire
> themselves when an enemy is detected. Science fiction writers can
> and do have a lot of fun imagining such things. (Gubrud 1997)

"Fertile grounds for fantasy" represent much of the hype surrounding nanotechnology today. Most such statements belong in the realm of science fiction. Discussions have included the concept of weapons that "fire themselves" when an enemy is detected. To the knowledge of this author, under all circumstances, decisions in the U.S. have avoided consideration of any such weapons; a decision to take a life is not left to a mechanical device, even a "smart" one involving the processing capability of a computer.

Another paragraph, representative of the profuse appearance of hype in the community, is:

> With nanotechnology, you can build a machine the complexity of
> a fighter jet the size of a gnat. If the aliens put just a few percent
> of the mass of machinery they are shown as having, in the form of
> military gnats, the humans are sunk. The gnats can be everywhere,
> not just one per city. What's more, the humans have nothing to
> shoot back at. They can protect themselves with hermetically
> sealed suits and buildings, but how many of us have those? The
> gnats simply fly up to you, inject a few micrograms of botulin
> toxin or the equivalent, and you become very extremely dead.
> (Storrs 1996)

Public perceptions are formed by press releases and interviews, which have echoed many irrational thoughts about this subject without any scientific validation. Today, exchanges on the Internet are beginning to have nearly as big an impact as other media. Viewing the subject of nanotechnology on the Internet today reveals a vast medium of hype and misperceptions. Any program associated with nanotechnology must be concerned about the implications of a surge of interest by young, impressionable students (mature adults are also included!) influenced by such distorted views.

Terrorism

Will the activities of nanotechnology be used in the future to add to the repertoire of terrorist activities? Can dangerous species be created willfully or by accident? Could research in this field provide a means of placing dangerous weapons in the hands of irresponsible individuals? Bill Joy, co-founder of Sun Microsystems, addresses this question in his initial article (Joy 2000). It is this point that is found to create the most concern by the many newspaper articles that followed his article. Consider the exchange appearing in the London Times:

> Q (by the interviewer): Is it a question of whether we should do something rather than whether we can do something?

> A (by Bill Joy): It's clear there are some things we shouldn't do. Genetically engineered viruses for example shouldn't be touched. But people can now give each other bioengineering equipment for Christmas. Clinton and Blair are trying to make technology available to everyone without first thinking about whether those people understand its uses. If we continue the way we're going, without thinking about the consequences, the technology will be misused. This technology can be used for genocide. You can't just sit by even when this word is hinted at. .....

> I don't think rogue corporations are a likelihood, it is much more likely to be smaller groups. Look at Monsanto; they were stopped without even losing a lawsuit. If corporations became more powerful then maybe... but the dangers I think are much more likely to come from terrorism and individual craziness.

This exchange suggests the greatest concern by the lay public relates to the ability of an individual or small group to knowingly or unknowingly create a species (in this case, biological in nature) that may introduce a menace in the form of a harmful virus or self-replicating entity unleashed in the environment (or society). Such attempts have been documented in Japan (Drell 1999).

Self-Replication

"Autonomous, self-reproducing machines are a computer-science quest that dates back to John von Neumann in the 1940s" (Storrs 1996). Artificial life is a subject of active investigation today (Di Paolo 2000; Lipson2000). It is safe to say that many variations in the DNA structure of viruses and bacteria have appeared through genetic mutations over the last millions of years. Many of these have wiped out large populations of living species (e.g., the Black Death). Individuals having the necessary genetic characteristics and immune systems to survive these viruses have passed this capability to succeeding generations; thus we are here today by virtue of our ancestors who have survived many diseases that have appeared throughout history. There is always the possibility, however, that a new variation or mutation may occur that could unleash harm to large numbers of individuals who have not or cannot develop the necessary defense mechanism.

A quote from the *San Francisco Chronicle* indicates the fear associated with self-replication:

> What deeply worries him [Bill Joy] is that these technologies collectively create the ability to unleash self-replicating, mutating, mechanical or biological plagues. These would be 'a replication attack in the physical world' comparable to the replication attack in the virtual world that recently caused the shutdowns of major commercial Web sites.

> 'If you can let something loose that can make more copies of itself,' Joy said in a telephone interview, 'it is very difficult to recall. It is as easy as eradicating all the mosquitoes: They are everywhere and make more of themselves. If attacked, they mutate and become immune.... That creates the possibility of empowering individuals for extreme evil. If we don't do anything, the risk is very high of one crazy person doing something very bad.' (*San Francisco Chronicle* 2000)

Consider the conditions necessary for self-replication. A virus is about the simplest self-replicating species that nature has evolved. It is constrained for nourishment to living species on which it may find the necessary nutrients and environment (e.g., temperature) for self-replication. Life as we know it consists of a rather complex array of chemical constituents. It is only through contact or some transfer mechanism (the wind in the case of plants) that a virus may be passed from one member of a living species to another. Viruses do not have the "intelligence" to create nutrients from non-living elements that they can then use to replicate; they require the rich array of nutrients found in living species. Further, viruses generally do not kill their hosts (Drell 1999), or else they would eliminate their means of sustenance. A fatal virus is typically one that is transferred from one species that has developed tolerance to another that has not.

A definitive answer does not exist today, but a reasonable hypothesis is that any self-replicating entity having the capability of sustaining itself by altering the environment (and gathering sustenance from non-living species) must be far more complex than a virus. Such an entity must be capable of gathering elements from various portions of the land, and in communicating among like entities (implying a social structure) to distribute these elements into whatever is necessary for sustenance and replication. Otherwise, a virus-type self-replicating entity will resemble the viruses we know today. This also requires a "resource" of living species that come in close proximity with one another in order for the virus to spread. In other words, self-replicating organisms are likely to resemble viruses as we know them today, or be very much more complex (and unlikely to be created except for years of advanced research far beyond any state of knowledge we have today).

The conclusion of the above paragraphs suggests that any concerns we have about self-replicating entities are likely to be restricted to those resembling the viruses we are familiar with today (and biological in nature). It is true that altering the DNA structure of a virus may produce a new species not heretofore created by the "random roll of the dice," and this is the reason for the caution expressed by Bill Joy: "Genetically engineered viruses for example shouldn't be touched." Genetic modifications that produce vaccines, however, may be considered advantageous if pursued with great care.

Taking Control

The issue of control is one of the major concerns of critics such as Bill Joy. This appears in the following (Chaudhry 2000):

> 'People are afraid precisely because there are no hurdles anymore,' Davis said. 'When you broaden the horizon far enough, there comes a point when what we know and what we can control drops away. This is very much about losing control.'

> 'Joy, however, is more worried about what he perceives as a refusal to take control of technology. He says scientists are taking a passive attitude toward technology, abdicating their moral responsibility to make responsible choices.'

> 'There is this fatalism,' he said. 'Like it's all going to happen anyway, and we can't do anything about it.'

If anything is needed, it is the presence of a responsible and respected scientific body with thoughtful statements about the reality of the options and possibilities arising from research in this area. The scientific enterprise has on occasion undertaken introspective examinations to improve public perceptions of the enterprise (Forum Proceedings of Sigma Xi 1993). Part of any analysis should include an accepted set of observations, which, if they begin to come true, represent a "signal" to give attention to developments that may represent danger as agreed upon by prior considerations. Due to the attitude expressed by many vocal participants in the debate about nanotechnology, serious discussion of self-regulation is probably due. Consider the viewpoint expressed in the following paragraphs:

> The proponents of bioelectronics are inevitably correct in suggesting that it holds out incredible benefits for the human race. (Admittedly, those who argue for human obsolescence as a benefit should be discounted by any reasonable humanist.) Likewise, it is undeniably the case that some of the skepticism toward bioelectronics arises out of the superstitious attitude that people hold toward computers and electronic technology, as well as medical and reproductive procedures that they don't fully understand. However, they are incorrect in arguing that regulation and oversight will only hinder research in this area and prevent scientific progress in the relevant areas. In marginalizing the social and ethical issues generated by research in biocomputing, these researchers are showing a side of science that people have routinely expressed anger about — its refusal to accept social responsibility for unforeseen consequences. In order for bioelectronic research to progress, it will have to accept that the potential dangers are real, and that the concerns of some skeptics are valid. Otherwise, something disastrous might occur which might create a "death-blow" for the industry, much as has happened with nuclear power in the U.S., and nothing positive will ever have been attained.) ... (Forum Proceedings of Sigma Xi, 1993)

> A new "cyborg bioethics" may be necessary. While it cannot be possible to foresee all the consequences resulting from bioelectronics, most scientists are already aware of what some of the major dangers are. Researchers in biocomputing may be required to adopt protocols on acceptable research with human subjects, much as genetic engineers did back in the 1970s. In drafting bioethical imperatives for bioelectronics research, it will probably be imperative to consider the concerns of groups such as the religious community, since to ignore their concerns simply out of the insistence that they are merely acting out of "anti-science" ignorance will leave an important group "out of the loop" of this research. This is uncharted territory for the human race, and it is the first time in which our own "built environment" may be directly incorporated into our own sense of self and human nature. Our own biocomputers (the human mind) evolved under a very specific set of evolutionary circumstances, after all, and they may

not be equipped with the foresight and moral sense to keep up with the accelerating pace of technology.

Since this is the case, it is probably imperative for society to assert that the scientists and engineers charged with creating this new technology exert the proper amount of social responsibility. Safeguards will have to be insisted on to prevent the possible negative impacts discussed above, and many of these things will have to be built in at the instrumental level, since they probably cannot be achieved only through policy and regulation. Critical public awareness and vigilance, of the kind already shown by Jeremy Rifkin and the Foundation on Economic Trends with regard to biotechnology, will be essential. But ultimately, bioethicists will have to grapple with the fundamental issues involved, which touch on aspects of human existence and human nature which reach to the core of what most people think is involved in what it means to be human, and this will not be an easy dilemma to resolve. (Mizrach, no date)

Other authors, who consider the present threat from biological agents, echo the point: "We're already there!" (Drell 2000). The concerns that have been expressed about self-replicating species from nanotechnology are very similar to those expressed about biological agents that can be made today. If steps are to be taken to control or regulate certain aspects of nanotechnology (genetics, nanotechnology and robotics), many of the lessons and concerns have already been extensively considered by those struggling with the threat of biological warfare (Drell 1999).

Rational Progress

It is imperative to view potential scientific and technological progress rationally. Constraints have been imposed on research involving cloning of human embryos or subjects. It is so difficult to envision the ramifications of cloning the human species that action was necessary to place restrictions on such activity. Laboratory demonstrations involving nanotechnology, at the present stage of development, do not even begin to approach the level of impact foreseen by cloning. A current dilemma is that "visionaries" who have not been involved with laboratory research foresee events that are far removed from the reality of what is possible given the current stage of research.

Further, consider the case of chemical/biological warfare. Biological warfare clearly involves the use of self-replicating species that destroy selected forms of life. The world has seen the consequences of chemical warfare, and, in limited scenarios, the use of biological agents (Drell 1999).

The horror resulting from the use of these agents has led to international agreements (Geneva Convention 1925; Washington, London, and Moscow Convention 1975). The use of such agents has been banned (agreed upon) by 142 nations (Crowe 1999), although today "there are over twenty countries with known or suspected chemical and biological weapons programs" (Mark 1999). Nanotechnology has been mentioned on occasion as a means of enhancing delivery of this threat (Hughes 1998) as well as a means of ultra sensitive detection in the presence of such a threat. Research into certain aspects of this form of warfare must continue:

> Work on offensive biological weapons is forbidden by law in the
> United States. However, the same is not true of many potential
> adversaries. Thus, it is important to have a vigorous research
> program to explore genetic mechanisms that can be applied to
> protecting our people from attacks using biological weapons.
> (Mark 1999)

The harm that may be caused by a "self-replicating" system is clearly one of major concern. U.S. Code*prohibits the possession or use of biological agents. Adversaries have considered a long list of such agents for use (Alibek 1999). Additional agents, be they biological or "self-replicating," would probably come under the same restrictions and controls. Self-replicating species evolving from genetic modification of DNA would almost certainly be considered as a biological entity.

A significant (but perhaps moot) question revolves around the possibility of creating a self-replicating species that is so dissimilar to those of biological

---

* United States Code Title 18, Part I, Ch. 10, Sec.175:

(a) In General. - Whoever knowingly develops, produces, stockpiles, transfers, acquires, retains, or possesses any biological agent, toxin, or delivery system for use as a weapon, or knowingly assists a foreign state or any organization to do so, or attempts, threatens, or conspires to do the same, shall be fined under this title or imprisoned for life or any term of years, or both. There is extraterritorial Federal jurisdiction over an offense under this section committed by or against a national of the United States.

(b) Definition. - For purposes of this section, the term "for use as a weapon" does not include the development, production, transfer, acquisition, retention, or possession of any biological agent, toxin, or delivery system for prophylactic, protective, or other peaceful purposes.

agents that new legislation or international protocols would be necessary. Software versions of viruses are clearly a problem today, and are sufficiently different from biological agents that legislation is necessary. However, viruses that consume materials and are not based on biological components are unlikely to be of concern for many years.

Semantics: Facts and/or Fiction?

Much of the concern that has been generated by Bill Joy and the publicity associated with research genetics, nanotechnology and robotics (GNR) is due to visions conjured by the use of words in an inappropriate context. Scientific principles are not changing as a result of nanotechnology. Some attention to translating current-day hype in terms of accepted science would bring reality to some of the science fiction that pollutes rational thought on this subject. It is unlikely that the Laws of Thermodynamics will be modified by nanotechnology or other scientific frontiers. If there were a hint that the laws of thermodynamics might be modified when one enters the nanometer regime, this would gain a great deal of attention by a large number of scientists for scientific validation.

Consider the "dictionary" associated with the words and phrases that appear in the non-scientific world (particularly on the Internet):

Molecular machines and/or assemblers: Such entities are envisioned to perform functions "atom by atom" to create products having "every atom in its place." In reality, such devices are little more than *catalysts* involved in a material transformation (reactants to products). Ordinary laws of thermodynamics will continue to provide the guidelines of what products are possible. Diamond-like products can be produced if the Gibbs free energy of the products is less than that of the reactants. Considerable effort by researchers over decades has found a few selected conditions where diamond products can be produced. It is very unlikely that such products will emerge from reactants in solution at room temperature, for example. Products that can be imagined are not necessarily easy to produce by reaction pathways.

Robotic life forms, living machines, self-evolving machines: Such terms conjure visions of self-replicating species that would be a threat to life as we know it today. Calling such entities "machines" invokes the same fears of a mechanized world that challenged "John Henry," the "steel-driving man," who "laid down his hammer and he died." Such self-replicating species, or "smart machines," have not appeared in any form other than unexpected viruses that have emerged from other living species. We can expect other self-replicating species such as new viruses. We are far from any

230

experimental evidence of other forms of self-replicating species. There should be ample time to address such a problem if any experiments begin to demonstrate effects that are science fiction today.

Self-replication: A self-replicating species is interpreted differently in various disciplines. In computer science, for example, an algorithm that is able to generate a sequence of bits representing an identical algorithm is considered to be self-replicating (Byl 1989). A computer virus is self-replicating. This ignores the hardware and energy provided to allow such an algorithm to execute. Such a "self-replicating" structure is very different from an assembly of atoms or molecules that constructs a replica of itself from "nutrients" available on earth. Semantic confusion persists when different disciplines attempt to communicate using such different preconceived concepts.

Gears: Molecular gears are envisioned rotating on "frictionless" bearings within components of a molecule. In fact, the exchange of energy between two components of a molecule through vibration-rotation interaction, particularly with proposed structures (and not produced experimentally) will have a high degree of interaction and energy exchange. They will provide a strong interaction between "moving components," and should not be envisioned as useful components of a "machine" until experimental evidence is obtained demonstrating that point. The laws of conservation of energy must be observed.

Molecular motors: Nature has provided living cells with remarkable structures having mobility and the ability to propel themselves. The term "molecular motors" has been used to label these entities. This is currently a subject of fascinating scientific research. The functions of these molecular motors are not completely understood. It is a leap of faith and imagination to assert that "molecular motors" will be used in a manufacturing process not under the influence of the laws of thermodynamics. There is much to be gained by research with these molecules. Science fiction, imagining bizarre consequences, should not alter valid scientific inquiry until *experimental evidence* begins to suggest processes that could be harmful.

Smart materials: Biological molecular structures (including viruses) have shown an amazing capability of selecting very specific forms of interaction with selected biological counterparts. These interactions can synthesize desirable products, or destroy a living cell. The term "smart materials" has been used in describing such molecules. A leap of faith deduction has led to concepts that such molecular structures could have extraordinary computer capabilities, and larger molecular structures could form the nucleus of an intelligent life form. It's true that an ant is a small living species with programmed behavior that exhibits even asocial behavior. An ant is based on biological principles, and does not violate any laws of thermodynamics. It has a rather limited brain, does not contain a very smart "computer," and is very limited in the behavior it may exhibit. Just how much "intelligence" can be contained in a given volume or mass of material is a question we don't know how to answer today. However, using the term "smart materials" for molecules with very selected functions should not be confused with "intelligent sophisticated computers."

Visions of "computers the size of a pinhead" have been propagated by members of the nanotechnology community to a lay public that does not understand the concept of smart materials. When the public hears these words, visions of machines (as they know them, with metallic gears, motors, etc.) coursing through the body cause great concern. Part of the problem has been the terminology used by "visionaries" in attempting to gain recognition for their efforts. These "visionaries" would be better employed by the filmmakers in Hollywood. Part of the problem is that institutions have been set up to further disseminate (or popularize) these views, to sponsor meetings, or even to attract venture capital. The nanotechnology community must be concerned with their image if Wall Street finds a lack of credibility associated with these commercial practices.

A responsible scientific community would be able to influence responsible scientists to use terms with specific meaning that are not emotionally loaded to please newspaper reporters. The mass media (newspapers, magazines, Internet Webmasters, etc.) have a responsibility to verify and validate statements made by alarmists. This should be pursued by the responsible community as part of an effort to reduce public concern over non-existing threats.

"The World is Coming to an End!"

Throughout history there has been a tendency for peripheral elements of society to feel that the world is coming to an end. The cartoon of a man in rags carrying such a sign is legend. Isaac Asimov has addressed many possible catastrophes leading to the end of humanity in his book, *A Choice of Catastrophes* (Asimov 1979). He would be amused at yet another variation to the many "choices" outlined in his book. A writing in 400 B.C. represents early concerns that have been with us as long as humanity has existed:

> Alas for the day! for the day of the Lord is at hand, and as a
> destruction from the Almighty shall it come. (Asimov 1979)

Critical Experiments or Observations

At the present stage of research in nanotechnology, little concern about "self-replicating life forms" exists among scientific investigators. Most of the fear expressed today comes from individuals influenced by a "virtual unreality" generated by "visionaries" who have taken free license to imagine both the best and the worst of what is conceivable (and not even possible). However, it is time to ask the question: At what point is it appropriate to express concern and for the government to develop guidelines over limiting research that may be potentially threatening to society? This subject may be discussed extensively.

A set of experiments, if demonstrated, could represent the stage at which social concern may be appropriate. The following set is offered as a beginning for discussions on this subject. Such a set should be examined and reformulated by responsible individuals until an appropriate set can be agreed upon. These can be reformulated as we learn more about the nature of the chemical, biological and physical world. Consider the following potential developments in nanotechnology. If in the course of nanotechnology research and development, laboratory experiments begin to reach the state indicated below, it is time to consider that such achievements may be subject to control if pursued to greater sophistication, in order to reduce the threat to society:

1. *Consumption of Resources:* Self-replicating species could possibly be produced that, if released, could uncontrollably consume resources required by a living species, or represent a threat to a living species outside of a laboratory. Note that computer viruses come under this same concern. Computer viruses consume information and time for individuals. The fact that computer viruses have been demonstrated represents a far greater threat than an imagined self-replicating robot that is only faintly conceivable in the distant future.

2. *Inadvertent Production of a Threat:* Self-replicating species (and this includes biological species) can be made that have a DNA sequence unlike that of existing species through a "roll of the dice" combination just to "try something different." This is particularly true of species that may resemble viruses, bacteria, or "life forms" that are known to represent threats to life. Such "random" experiments with new forms of living species should not take place.

3. *Computing Machines No Longer Responding to Humans with Programmed Predictability:* Assume that computer or logic functions

can be made whereupon such computers are no longer completely responsive to human control (recall the computer "Hal" in the film "2001"). Such systems would be considered inappropriate for design or production. This is not meant to preclude computers with artificially intelligent algorithms, but rather to computing machines that have a "mind of their own."

4. *Devices Lulling Humans Into Acquiescence:* Any combination of computers, robots, and self-replicating species that appear to take over human functions and simultaneously lull human activity into acquiescence (or a subordinate roll) should be considered a threat. Some have suggested that television already falls in this category. A level of "control" by a device is the main issue to be dealt with.

5. *Inexpensive Products Used to Unduly Influence:* Any material, device or organism that can be used to "unduly" or "illegally" influence one individual, group or nation over another should be considered a threat to society. Weapons of mass destruction (nuclear, chemical and biological) come under this category. Mass indoctrination is another. That is why such overwhelming attention is given to these weapons. Materials, devices or organisms that may be fabricated and used by terrorists to influence others come under this same category. If small groups can inexpensively produce sophisticated products having "undue or illegal influence," this becomes a subject for attention and potential legal action and/or restrictions.

Recommendations

1. *Enhanced Defense Systems:* National Defense will be significantly enhanced by a nanotechnology S&T program. Many aspects of current high technology defense systems and procedures are envisioned to have improved capabilities with the anticipated products of this field. Advanced technology is a key element of our national security; we therefore must pursue this subject vigorously. As a means of emphasizing the most relevant aspects of nanotechnology S&T for this purpose, strong support within laboratories emphasizing national defense missions is most appropriate.

2. *Address Integrity of Nanotechnology:* The nanotechnology community should, in an appropriate forum, address the misinformation about the subject that appears in the popular press and the Internet. A lack of scientific discipline associated with many hypothetical products of nanotechnology can negatively impact the integrity of the science and the image of the field. This forum should address the hyperactive

misperceptions about self-replicating species. Issues should be recognized that might be different (if any) from those already faced by the current threats from biological agents.

3. *Address Societal Impact of Nanotechnology:* An appropriate forum should address the potential impact of the anticipated products of nanotechnology on society. This should take the form of searching for agreement on a set of experiments or observations which, if found to be true, would represent capabilities that are not in the best interests of society. These observations and issues should include ethical and moral as well as threat questions.

4. *Distributed Resources Lead to Greater National Security:* To the extent that nanotechnology provides enhanced resources in the form of (1) new systems, (2) increased capabilities of existing systems, or (3) reduced costs for the performance of existing capabilities, this represents an increase in available resources for the world. With increasing resources distributed worldwide, tensions between nations and groups tend to be less, resulting in enhanced national security for all. It is also noted that nanotechnology is being pursued vigorously worldwide, enhancing the opportunity for worldwide distribution of the benefits of research in this field.

References

Alibek, K. A., Testimony of Dr. Kenneth Alibek, Chief Scientist at Hadron, Inc., before the House Armed Services Committee, October 20, 1999. Ref: http://www.house.gov/hasc/testimony/106thcongress/99-10-20alibek.htm.

Asimov, I., *A Choice of Catastrophes*, Simon and Schuster, NY (1979).

Byl, J., Physica D 34,295-299 (1989). *Self-Reproduction in Small Cellular Automata.*

Chaudhry, L., *Valley to Bill Joy: 'Zzzzzz'*, WiredNews, Apr. 5, 2000. Ref: http://piglet.ex.ac.uk/mail/cybersociety.2000/0305.html.

Crowe, W. J., Statement of Admiral William J. Crowe, Jr., before the House Committee on Government Reform, October 12, 1999. Ref: http://www.house.gov/reform/hearings/healthcare/99.10.12/ crowe.htm.

Di Paolo, E. A., *Artificial Life Bibliography of On-line Publications*, June 28, 2000. Ref:http://www.cogs.susx.ac.uk/users/ezequiel/alife-page/alife1.html.

Drell, S. D., A. D. Sofaaer and G. D. Wilson, *The New Terror, Facing the Threat of Biological and Chemical Weapons*, Hoover Institution Press, Stanford University (1999).

Drell, S. D., A. D. Sofaaer, and G. D. Wilson, Hoover Digest 2000, No. 1, *The Present Threat*. Ref: http://www-hoover.stanford.edu/publications/digest/oo1/drell.html.

Forum Proceedings of Sigma Xi, the Scientific Research Society, *Ethics, Values, and the Promise of Science*, Feb. 25-26, 1993.

Geneva Convention, *Protocol for the Prohibition of the Use in War of Asphyxiating, Poisonous or Other Gases, and of Bacteriological Methods of Warfare*, signed at Geneva on June 17,1925.

Gubrud, M. A., *Nanotechnology and International Security*, Fifth Foresight Conference on Molecular Nanotechnology, November 5-8, 1997. Ref: http://www.foresight.org/Conferences/MNT05/Papers/Gubrud/.

Hughes, P. M., *Global Threats and Challenges: The Decades Ahead*, Lt. Gen. Patrick M. Hughes, Director of the Defense Intelligence Agency, in Statement for the Senate Committee on Intelligence, January 28, 1998. Ref:http://www.fas.org/irp/congress/1998_hr/s980128h.htm.

Joy, Bill, *Why the future doesn't need us*, http://www.wired.com/wired/archive/8.04/joy_pr.html.

Kennedy, P., *The Rise and Fall of the Great Powers, Economic Change and Military Conflict from 1500 to 2000*,Random House (1987).

Lipson, H. and J. B. Pollack, Nature, 406, 974-978, (2000). *Automatic design and manufacture of robotic life forms.* See, also, several additional articles on this subject in contemporary issues of Nature.

Mark, H., Statement of The Honorable Hans Mark, Director, Defense Research and Engineering, before the House Armed Service Committee, October 20, 1999. Ref: http://www.house.gov/hasc/testimony/106thcongress/99-10-20mark.htm.

Mizrach, S., *Should there be a limit placed on the integration of humans and computers and electronic technology?* Ref: http://www.limmat.ch/koni/texte/cyborg-ethics.html.

Roco, M. C., R. S. Williams, and P. Alivisatos, *Nanotechnology Research Directions: Vision for Nanotechnology in the Next Decade*, Kluwer Academic Publishers (2000).

San Francisco Chronicle, *Scientist Is Fearful of Computer Mutiny; Sun Micro co-founder says replicating robots could replace humans*, March 13, 2000.

Sterling, B., *When Robots Act Like Rabbits*, The Wall Street Journal, September 6, 2000.

Storrs, J., Hall, *Flying Saucers?*
http://www.nanocentral.com/NanoWorld/Perspectives/JoshHall/askjosh_aug96.html.

Washington, London, and Moscow Convention, Convention on the prohibition of the development, production and stockpiling of bacteriological (biological) and toxin weapons and on their destruction, signed Washington, London, and Moscow April 10,1972, entered into force March 26, 1975.

6.5 FOCUS ON SOCIAL, ETHICAL, LEGAL, INTERNATIONAL AND NATIONAL SECURITY IMPLICATIONS

SOCIAL SCIENCE RESEARCH METHODS FOR ASSESSING SOCIETAL IMPLICATIONS OF NANOTECHNOLOGY

J.S. Carroll, MIT Sloan School of Management

Scientific discoveries by themselves rarely create change. It is the confluence of old and new technologies with old and emerging social needs that creates change. There are dynamic interactions and "tipping points" that are difficult to foresee (e.g., the capabilities that emerge when computers get one more order of magnitude faster, coupled with a clever new idea for software, leveraged by increasingly computer-literate 20-year-olds). Thus, the problem for prediction is not only to track the advances of nanotechnology, but also to track other advances and changes in society at the same time. Visionaries and science fiction writers have bold visions, but researchers must also take a role in producing a useful understanding of possible societal directions.

Perhaps the best we can do is to open our thinking and be more aware of the kinds of changes that may occur, how to spot them as early as possible, and how to prepare to influence the course of changes when deemed necessary or desirable. Of course, it is important to consider who would be doing the "deeming." History suggests that those in power tend to suppress or co-opt new technologies (e.g., RCA's success in subverting Farnsworth's patents on TV and convincing the public that they were the inventor, Fisher and Fisher 1997), and the newly empowered try to do the same (cf., Bill Gates).

So, I accept my more modest role to suggest social science research methods for studying societal implications. I begin with some assumptions about research and nanotechnology, and then present goals for social science approaches to societal implications of nanotechnology. I then suggest sources of measures and kinds of indicators that can be helpful, including "leading indicators" that might provide early hints of change. Finally, I offer possibilities for types of research designs and some conclusions.

Some Assumptions

Nanotechnology, as a family of tools and techniques, is a source of products and other "stuff." It is these products and stuff that will be fought over, demanded, and used in ways that will bring layers of change. Each of these

238

bits has the potential to be mundane variations with a bit more zip or a bit less cost, or to be dramatic advances. We are not likely to know which is which, and the nanotech equivalent of the hoola hoop may turn out to be critically important when combined with something else.

Initially, the impacts of nanotechnology will be via specific products and innovations. Such primary effects would be to make things work better, cheaper, with more features, etc. This might, for example, increase food yields, generate new textiles for clothing, improve power production, cure certain diseases, or whatever. Secondary effects might be shifts in demand for products and services, so that people come to expect different kinds of food, medical care, entertainment, etc. The required infrastructure for nanotechnology may create interdisciplinary research centers, new educational programs to supply nanoscientists and nanotechnologists, etc. Later, tertiary effects would move upstream in our social structures and cultural patterns, such as shifts in education and career patterns, family life, government structure, and so forth[7]. Will nanotechnology extend our lifetimes more rapidly than it extends our health, or vice versa? Will nanotechnology enable so much connectivity to information and to each other that it revolutionizes society (cf., Star Trek's The Borg) or shift us more toward solitary lifestyles of "distance experiences" and "virtual experiences" rather than personal contacts (cf., Asimov's *The Naked Sun*)?

Research, including social science research, is an activity intended to generate and validate knowledge, based on systematic rules agreed to by a community of scientists. Since there are many subcommunities of scientists, the rules vary somewhat from discipline to discipline, place to place, and time to time. It is important to realize that the activities of social science research also influence policymakers and the general public, and thereby change the way society thinks and acts (Gergen 1973). To measure

---

[7] Imagine, for example, the impact of an increased ability to sequence the human genome and identify consequences of various genes. A practice of "genometrics" could spring up that initially ran new tests on at-risk populations and cured some diseases. Over time, people might begin to use such information first for their own curiosity and later to select their workers, spouses, or children. Companies might spring up to give genetic advice and private schools might begin tailoring education to genetic codes. Socially, it might become fashionable to "wear your code" (or those aspects of your code that offer social status) and political candidates might be forced by public opinion or law to disclose their codes. Ultimately, society might be reorganized into genetic classes, as starkly portrayed in the movie Gattaca and earlier in Huxley's *Brave New World*.

something is also to shape the future, by changing what we pay attention to, expect, reward, and punish. Therefore, it is important to measure both what is desired, and also what is feared.

Research Goals

In light of the above assumptions, it seems reasonable to pursue the following goals for social science research on the societal impacts of nanotechnology. First, we need to define and measure "societal impacts." Second, we need to find leading indicators or first signs of impacts. Third, we need to develop theories that explain impacts, identify causal mechanisms and contingent conditions (e.g., under what circumstances would particular products have particular impacts), relate various advances and impacts together in more comprehensive systems models, and permit (tentative) extrapolation to possible futures. Finally, we would like to assist policy development on the basis of what is known from our research and what is known about desires and values, i.e., what are "society's" goals and how will these goals change over time as technology advances?

Measures and Indicators of Societal Impacts

There are a huge number of potential measures for societal impacts. Our standard measures of the social world are all useful and all insufficient. From a methodological viewpoint, there are various techniques for measuring societal phenomena, including self-report surveys and interviews; diary studies (e.g., new software that allows Palm Pilots to interview their owners) and think-aloud protocols; social and economic statistics; direct observation (of, for example, meetings, point of sale, point of use); bibliometric and content analysis (of patents, citations, email, Web sites and hits, chat room content, 10Ks); network analysis of relationships across people and organizations; accretion measures of what piles up (e.g., trash heaps); and erosion measures of what gets worn out (e.g., repair calls). For some general principles and examples, see Babbie 1989; Judd et al. 1991; Webb et al. 1981).

It may be helpful to sort through potential measures by using some conceptual criteria or categories. For example, we could distinguish *process* from *content*. The temporal *process* might include scientific or social visionaries who first imagine possibilities, nanoscience and nanotechnology discoveries that connect with these possibilities, actual products that are developed and marketed, public acceptance by scientists, consumers, and policymakers, substitutions of usage patterns as new products replace old products, interactions of new product capabilities with existing technical and social arrangements that reshape demand and usage, and finally transformations of social institutions and associated infrastructures. The *content* domains might include health, wealth, food, fuel, productivity, education, employment, national security, happiness, social capital (networks of relationships), political participation, ethical thought, etc.

From the process categories, we can generate more specific indicators of innovation and change. Visions are likely to appear in the media, at meetings, and on the internet. Discoveries can be tracked through patent applications and new products and services. Investments are visible in budgets, grants, strategic plans, projects, job titles, hiring results, and educational programs. Public acceptance is indicated by attitude measures, purchases, usage, and even charitable contributions. Interactions show up in studies of professional and other social networks, fields of study, and communities. Substitutions are evident in the decline of old industries. Transformations can be measured in indicators of lifestyle and social institutions such as deurbanization and demassification (see Brown and Duguid, this volume).

Other conceptual criteria underlie the objectives and recommendations in the IWGN Workshop report. These constitute a nascent "theory" of the nanotechnology process and impacts. The report divides its objectives and recommendations into sectors or classes of participants such as academe, private sector, government labs, funding agencies, and professions. For each sector, the objectives suggest process and/or content measures of change. In academe, the workshop endorsed interdisciplinary work, new courses, fellowships, information flow, and regional coalitions. For the private sector, the focus was on investments, startups, and coalitions. For government labs, the report looked at budgets, equipment, standards, and coalitions. For funding agencies, the priorities were on new initiatives, databases, and centers. The professions were expected to create new forums, symposia, and job fairs where interdisciplinary topics and careers could flourish.

Of particular importance would be indicators that provide early signs of change. In our process indicators, we would probably look upstream to the beginning of the chain of events and examine allocations of effort in academic and corporate R&D labs and analyses of patent applications to see the emerging technological trends. More generally, we might consider who, what, and where we could find "bellwethers" or first movers. California seems to be the first place that social change occurs in the United States. Finland and Singapore are examples of countries that have embraced new technologies and undergone rapid change. Within both rapidly changing and slowly changing societies, there may be classes of "early changers" such as 15 year olds who are at the forefront. In some cases, such as in the health care domain, it may be the 90-year-olds who show the first signs of change. Science enthusiasts, nerds, and hackers may be first movers in a nanotechnology-rich world. Start-up companies, university labs, and Internet chat rooms may be places to look for changes and impacts. Particular social strata, such as groups "on the margin" of society, may also exhibit the early signs of changes (things that seem weird or bad) that will later spread to mainstream society.

Research Design

Research is more effective when it is designed into the process being studied, rather than having to explain what happened after the fact. It is not surprising that many of our best studies of the impact of technologies are retrospective analyses from 10 or 100 years ago (e.g., Bijker 1995; Fischer 1992) but we can't afford to wait that long for knowledge of current developments and impacts. If social scientists can be introduced earlier into partnerships and collaborations with nanoscientists and nanotechnologists, there is a better chance to learn more and learn quickly enough to guide policy. Research designs are typically stronger on measures and mechanisms if introduced early, so that measures can be made over time and informative controls can be established to strengthen causal understanding.

I have listed some typical design types in the order in which I would guess they will be used. In other words, designs that are easier to execute are likely to be used first, and those that require great skill and/or great control over circumstances may come later. Surveys, expert panels, and statistical summaries of socioeconomic data are relatively easy to do and provide useful snapshots. Simulations (computer-based or game-like) allow us to project behavior in hypothetical scenarios (e.g., Sterman, 1989). Case studies (Yin 1989) permit a rich set of information gathered around particular cases of inventions, products, companies, communities, supply chains, etc. Quasi-experiments (Cook and Campbell 1979) allow focused comparisons that strengthen causal inference, such as time series and econometric designs and control groups based upon diffusions of new technologies (some groups naturally get it earlier, some later). Ethnographies (Van Maanen 1988) and in-depth immersion in real-world sites allow a rich understanding of complex interdependencies and subtle phenomena. True experiments allow us to answer precise questions, but are difficult to arrange in real-world settings.

Conclusions

In conclusion, we must remember that we cannot easily predict the products and impacts of nanotechnology on society. We are not studying a monolithic "nanotechnology" but rather a host of varied technologies, products, services, and other interventions. Some implications and impacts are relatively easy to predict and to study, but others will be emergent and surprising. However, everything in society will be changing (not just what nanotechnology has touched directly), and everything will be connected together (loosely or tightly), so predictions will be uncertain and causal explanations will be difficult to validate.

Research is needed to help us understand changes and to plan action. A wide range of indicators will be needed because we do not know what will emerge or what will turn out to be important. Helpful theories are especially important to focus attention on key issues and processes, to guide research, and to represent the results of research. Society needs theories and system models to understand how changes in one part of the system, whether a particular type of technology or a particular element of society, spread out to create intended and unintended effects throughout the system. This includes understanding the impact of society on technology. This will enhance our ability to educate everyone earlier and to improve the quality of public debate. It will also enable scenario analyses, strategic planning, and simulated public debates to be more informative and rich.

At its best, the research attitude is one of openness, curiosity, sharing, and constant improvement. This is a model for increasing the capability of the public and the scientific communities to plan intelligently, communicate effectively, respond to emergent circumstances, and understand themselves and the broader society and world in which they live and work. From an ecological perspective, there is no guarantee that "progress" (however defined) has any particular consequences for the human race; nor is there a guarantee that more science and technology will always find an answer to human problems. We must improve ourselves as a thoughtful and ethical human society at the same time that we improve our mastery over the physical world.

References

Asimov, I. *The Naked Sun*. Garden City, NY: Doubleday, 1957.

Babbie, E. *The Practice of Social Research*, 5th ed. Belmont, CA: Wadsworth, 1989.

Bijker, W. E. *Of Bicycles, Bakelites, and Bulbs: Toward a Theory of Sociotechnical Change*. Cambridge, MA: MIT Press, 1995.

Brown, J. S. and Duguid, P. Don't Count Society Out: A Response to Bill Joy. Paper prepared for the Societal Implications of Nanotechnology Conference, 2000.

Cook, T. K. and Campbell, D. T. *Quasi-Experimentation*. Rand-McNally, 1979

Fischer, C. S. *America Calling: A Social History of the Telephone to 1940*. Berkeley: U. of California, 1992.

Fisher, D. E. and Fisher, M. J. *Tube: The Invention of Television*. Harcourt Brace, 1997.

Gergen, K. J. Social psychology as history. *Journal of Personality and Social Psychology*, 1973, **26**, 309-20.

Huxley, A. *Brave New World*. New York: Harper and Row, 1946.

Judd, C., Smith, E. and Kidder, L. *Research Methods in Social Relations*, (6th Edition). Holt, Rinehart, Winston, 1991.

Sterman, J. D. (1989b). Modeling Managerial Behavior: Misperceptions of Feedback in a Dynamic Decision Making Experiment. *Management Science*, 35(3), 321-339.

Van Maanen, J. *Tales of the Field*. Chicago: U. Chicago, 1988.

Webb, E. J., Campbell, D. T., Schwartz, R. D., Sechrest, L., and Grove, J. B. *Unobtrusive Measures: Nonreactive Research in the social Sciences* (2nd ed.). Boston: Houghton Mifflin, 1981.

Yin, R. K. *Case Study Research* (2nd ed.). Newbury Park, CA: Sage, 1989.

ETHICAL ISSUES IN NANOTECHNOLOGY

V. Weil, Illinois Institute of Technology

Because we are concerned with issues of practical and professional ethics, not theoretical ethics, we must examine nano initiatives along with associated social practices, institutions, organizations, and the choices and actions of individuals within them. In nano science and engineering, ethical and responsibility issues connected with various initiatives will reflect many complexities — those from the multidisciplinary and multi-institutional character of research and development, as well as from more technical aspects.

To survey ethical implications and identify risks it is necessary to step back from captivating visions of profound transformations of the material world and of society. The focus must be on specific proposed initiatives in their institutional environments. Normative questions, that is, questions about what it is right and appropriate to do, gain a foothold at the macro level in relation to societal and organizational policies and activities and, at the micro level, in relation to ethical standards, responsibilities, decisions, and actions of individuals.

An important aim of ethical investigation is to anticipate ethical problems -- preventable harms, conflicts about justice and fairness, and issues concerning respect for persons likely to arise from specific nano initiatives. A second important aim is to foster sensitivity to ethical issues and responsibilities at every level of decision making by both technical and policy people.

How to Proceed

At the outset, it is essential to clarify the term "nanotechnology." There is no disagreement about the oversimplifying and misleading character of that term. It is a catch all that has caught on because of its convenience and market appeal, its usefulness as a rallying point. Though aware of its deficiencies, interested parties are not inclined to reject use of the term. It is therefore necessary to make clear what are the initiatives, disciplines, and institutions that "nanotechnology" embraces.

(1) Initiatives with implications for advances in medicine, computing, space exploration, energy conversion and storage, optics, and materials, including catalysts, are among those often cited. (2) A large number of disciplines — and specialties within them — in science, engineering, mathematics, and computing are encompassed. (3) Institutions of government, academe, and industry, and organizations and practices within these institutions are also included. Government agencies act as promoters of initiatives, and they have begun to fund university research and graduate training. Universities are gearing up for and launching new programs, instituting graduate studies, renovating and putting up buildings, and forming alliances with other universities, national laboratories, and private firms. New collaborative arrangements between universities and private companies to carry forward research and development are being, and will continue to be, forged. Of use in forming these new collaborations is the experience recently gained from crafting such arrangements to advance work in information technology and biotechnology.

That experience suggests a second task: to examine recent history. Study of our experience with biotechnology and information technology may help to locate nodes of ethical concern. Caution in drawing parallels is necessary, however, in light of claims made for the uniqueness of nanoscience and nanotechnology. It may, nevertheless, be useful to direct attention, from an ethical perspective, to questions that these earlier technologies suggest. Included are questions about how government plays a role in promoting, launching, and supporting technological developments and about how projects to produce products are selected and by whom. Government agencies' promotional discourse to boost the NNI is already available for scrutiny, and there may be opportunities to study how nano projects are currently selected. Our history with earlier technologies suggests the need to devise processes and settings for information exchange with and wider participation by members of the public in order to promote transparency. (See Wynne 1991 for an especially insightful discussion.) So compelling are the ethical and practical benefits of building in openness, disclosure, and public participation from the outset that efforts towards those ends should begin without delay.

Recent history with other technologies indicates that the obstacles to achieving these aims are formidable. Observers have already expressed concern that multidisciplinary meetings present special risks to open exchange of information (NSTC 2000, p. 32). Similar concern is appropriate regarding the flow of information across institutional boundaries, between academe, government, and the private sector (Blumenthal 1992). Some problematic patterns relating to the flow of information between those engaged in technological development and the public are familiar from technological developments of the recent past. For example, by becoming locked into a definition of the public as the "other" — the enemy, uncomprehending, standing in the way of advance — those with authority over information may withhold it and thereby cut themselves off from public reactions. Failing to nourish genuine information exchange, they may invite the very opposition they wish to ward off. Such patterns from the past point to the need for care and caution in framing "the communication problem", to avoid seeing it as a problem of one-way communication downward (Wynne 1991).

A third task at this stage is to try to identify ethical issues that have already arisen or are likely to arise. As elsewhere in practical and professional ethics, it is necessary to disaggregate, to look case by case at specific nano science and technology options and their consequences — current, foreseen, foreseeable, or speculative. Initiatives and options should not be divorced from their institutional contexts, however complex the latter may be. At

246

times technical people fail to foresee what is foreseeable and within their sphere of responsibility. Sometimes they claim to foresee what is not foreseeable, given available knowledge. Nanotechnology options afford opportunities for both kinds of foreseeability problems to arise regarding the consequences of specific developments.

The recent history of rapidly developing technologies suggests that we should be alert to unintended consequences. Consider, for instance, the complex issues that have arisen about privacy in connection with information technologies (Johnson and Nissenbaum 1995). The privacy debates also illustrate how ethical issues concerning respect for persons are generated with the propagation of new technologies. Biotechnology offers another cautionary example relating to unintended consequences. When investigators in the new biotechnology achieved confidence in control of their products in the laboratory, vigilance regarding unintended consequences did not extend to the new products in the field. Information about interactions of new bioengineered organisms with other members of an actual ecosystem has been scarce and slow to appear. Accordingly, concern about unintended consequences from agricultural products of biotechnology remains high in many places (Weil 1996).

While trying to stay alert to unintended consequences, we should also try to avoid taking it for granted that there is wide agreement on the desirable consequences of various nanotechnology options. It is essential to obtain a diversity of perspectives on the desirability of particular options. For eliciting a range of perceptions, conversation between people in nano research and development and members of the public is necessary. It may be essential to create mechanisms, such as citizen/scientist panels, to bring different perspectives into the conversation. In the exchanges, people will have to learn to identify interests that are in play, their own and those of other parties to the conversation. Experience with biotechnology shows the costliness of proceeding with mistaken assumptions about what are desirable outcomes and products (Crow 2001).

Identifying Ethical Issues

To go further in identifying ethical issues, we need concrete points of departure, actual examples or cases that pose ethical questions, quandaries, or conflicts. This is how we proceed in practical and professional ethics when looking at issues in other social practices. Lacking specific nano examples at this time, we may note features of nanotechnology that, because of their alleged novelty, may be sources of ethical concern. For example, forecasts of development of new catalysts stress the creation of new production processes. The latter are likely to raise ethical questions, for instance, about the need for safeguards for workers that specific new processes might generate. Questions of these kinds can provide points of entry to the institutional, organizational settings in which potential problems are embedded and in which they must be examined.

The new processes are confidently predicted to produce workforce changes (NSTC 2000, 20). Anticipated impacts on the "human resource infrastructure" will surely bring benefits to some and harms to others. When there are winners and losers, issues about equity cannot be avoided. Again, on a case by case basis, it will be important to identify winners and losers so as not to inflict preventable harms or lose opportunities for mitigating harms or compensating losers.

Ethical issues associated with intellectual property protection are virtually certain to arise (NSTC 2000, 31). They will be novel insofar as novel features of nanotechnologies and their social environments introduce new complexities relating to intellectual property protection. In environments where patents and trade secrets are generated, there will be implications for open exchange among technical people and communication with the public. The rationales justifying ownership are likely to be as vigorously debated and contested as those associated with information technology and biotechnology. Equity issues raised by intellectual property protection should generate debate as well. Vigorous and extensive public discussion could even lead to reexamination and revisions of intellectual property policies.

Ethical questions about university/industry relationships are hardly novel, but they are virtually certain to arise. These questions have engendered discussion, literature, and large-scale empirical studies since at least the mid 1980s (Blumenthal et al. 1986). By now, some specialists contend that institutional accommodations to new relationships with private companies have transformed universities, bringing significant changes in university values and practices (Webster and Etzkowitz 1991). For instance, many in universities now accept a need to allow faculty members to accumulate great wealth through their research (Forest 2000). Yet there is a clear understanding on both sides that universities and private sector enterprises are valuable to each other as partners precisely because of their differences (Weil 1988; Weil 2000).

The close association of university research with the private sector has brought problems of conflict of interest to the forefront. For example, questions arise about whether a university researcher's ties to a for-profit firm threaten reliable judgment in university research. Observers have suggested that universities as institutions can have conflicts of interests (Frankel 1996; Pritchard 1996). A strong program of research and development in nanoscience and nanotechnology will subject university values and practices to new pressures, and universities will have to continue to make accommodations preserving core university values. They will have great need for ethical guidelines as they make those accommodations and review their policies in the light of experience.

The focus on institutions, organizations, and practices should not obscure the need to focus on the individual responsibility of engineers, scientists, and others involved in the processes of producing new technologies "with unprecedented control over the material world". To focus on individual responsibility, university programs of graduate study and research in nano areas should include attention to ethical issues specific to their own nano areas. They should give attention to these issues in their training of students in scientific research ethics and in their technical training of students. Professional societies have a role to play in affording opportunities for debate and discussion and helping to devise, for individuals and organizations, guidelines that incorporate ethical principles responsive to emerging issues.

Addressing the Issues

This survey of ethical dimensions suggests three main fronts of activity for responding to issues. One is activity within research and development initiatives funded by government agencies and carried out in universities. Government agencies should require an ethics component in grants for graduate training and in grants for research. In this way, they can provide for attention to ethical issues by individual researchers and principal investigators within research groups, by research groups as collectives, and by university departments and centers in which the nano research and training goes on. It is necessary to connect specialists in ethics and behavioral sciences with such projects from their outset and maintain the specialists' association with ongoing nano projects of research and development. National Science Foundation proposal guidelines already incorporate provisions requesting an ethics component. Such provisions should be standard components of requirements for submitting proposals in the nano domain to government agencies. Over time, the provisions can be further detailed or improved in other ways, as experience indicates.

The second front of activity has already been suggested -- a fresh and energetic effort to devise opportunities for genuine conversation with members of the public about current and proposed initiatives. In view of the ethical and practical reasons for commitment to disclosure and incorporation of democratic processes in advancing nano initiatives, government agencies that are promoting nanoscience and nanotechnology should accept an obligation to create appropriate channels and fora. Concerted activity on these two fronts could be innovative and capable of transforming research and development in step with the innovative and transforming features of nano processes, structures, and products.

A third front of activity is education. While some ethics specialists are already available to help in addressing the ethics component in research, development, and education related to nano initiatives, there are not enough. Mainstream graduate training in philosophy and behavioral science is not yet oriented to respond to this need. The intellectual interest and acknowledged importance of issues raised by the NNI justifies government initiatives to address the lack of qualified ethics and behavioral science specialists.

It should be possible to devise training programs led by existing specialists in collaboration with faculty who direct graduate and postgraduate study in appropriate disciplines. Postdoctoral studies may be especially suitable for providing the training and forming the collaborations across disciplines that are needed. A program to train people for the ethics work within nanotechnology and nanoscience initiatives is essential, and feasible, with government support. This education should be part of a concerted endeavor to improve education at all levels, in scientific and engineering disciplines, and areas that cross disciplines, as well as in philosophy, ethics, and behavioral sciences. For the NNI to begin to fulfill its promise of carrying forward research and development initiatives that bring benefits to society fairly distributed, concerted efforts on all three fronts are needed.

References

Blumenthal, D. 1992. "Academic-Industry Relationships: Extent, Consequences, and Management," *The Journal of the American Medical Association*, 268:3, December, pp. 3344 -3349.

Blumenthal, D.M., M. Gluck, K.S. Louis, M. Stoto, and D. Wise. 1986. "University-Industry Research Relationships in Biotechnology: Implications for the University," *Science* 232: pp. 1361-1366 (June 13).

Crow, M.M. 2001, "Harnessing Science to Benefit Society," *The Chronicle of Higher Education, Chronicle.com.* 9 March, 2001: 17 pars. March 8, 2001 (http://chronicle.com/weekly/v471/i26/26b02001.htm).

Forest, G. 2000. Comment at Conference on Societal Implications of Nanoscience and Nanotechnology, September 29, 2000, NSF, Arlington, VA.

Frankel, M. 1996. "Perception, Reality, and the Political Context of Conflict of Interest in University-Industry Relationships," *Academic Medicine*, 71:12, December, pp. 1297-1304.

Johnson, D.G. and H. Nissenbaum (eds). 1995. *Computers, Ethics and Social Values*. Prentice Hall:Englewood Cliffs, NJ. pp. 262-393.

National Science and Technology Council (NSTC). 2000. *National Nanotechnology Initiative: Leading to the Next Industrial Revolution*. February. (http://www.nano.gov/nni.pdf).

Pritchard, M.S. 1996. "Conflicts of Interest: Conceptual and Normative Issues," *Academic Medicine*, 71:12, December, pp. 1305-1313.

Webster, A. and H. Etzkowitz. 1991. "Academic-industry Relations: The Second Academic Revolution?" Science Policy Support Group, London, 1991.

Weil, V. 1988. "Afterword," *Biotechnology: Professional Issues and Social Concerns*. (AAAS Publication) co-edited with Paul DeForest, Mark Frankel, and Jeanne Poindexter.

Weil, V. 1996. "Biotechnology: Societal Impact and Quandaries," in *Biotechnology and Ethics: A Blueprint for the Future, Report of NSF Workshop*, Northwestern University, Center for Biotechnology.

Weil, V. 2000. "Como Pueden Convivir Los Valores Universitarios Y Las Normes Empresariales?" in *Dos Ejes en La Vinculacion de Las Universidades a la Produccion*. ed. by Rosalba Casas and Giovanna Valenti. Plaza y Valdes, S.A. de C.V. Instituto de Investigaciones Sociales. Universidad Nacional Autonoma de Mexico. Universidad Autonoma Metropolitana, pp.81-93.

Wynne, B. 1991. "Knowledges in Context." *Science, Technology, and Human Values*, 16:1, Winter 1991, pp. 111-121.

SOCIAL ACCEPTANCE OF NANOTECHNOLOGY

P.B. Thompson, Purdue University

The notion of "social acceptance" of technology is prevalent throughout both scholarly and social science studies of technology and in popular literature, yet it is, in an obvious way, a very obscure idea. On the one hand, it connotes empirical content, perhaps even measurable criteria, so that whether or not a technology has been socially accepted appears to be decidable question, a matter of fact about social relations or how things stand in the world. On the other hand, the phrase "social acceptability" suggests a normative judgment in a way that makes social acceptance come to involve inherently contentious characterizations of "society's values." Here I want to make a few general remarks about the social acceptance of new technology, to note some lessons learned and yet to be learned from the ongoing saga of biotechnology, and to make some suggestions about how these lessons might apply to nanotechnology.

Here are some of the indices that might be taken to measure social acceptance of technology.

- *Geographical Parameters*: Where is the technology used — and "where" can be characterized in terms of location, demography, culture, class, etc.

- *Economic Parameters*: What is the market penetration for the technology (what percentage of the potential purchasers of the technology actually purchase it), what is the price sensitivity, etc.

- *Psycho-Social Parameters*: What do surveys indicate when people are asked how what they think about a given technology?

- *Affective Parameters*: What's the "comfort level" of users? Do they feel a sense of regret? Do they feel a sense of moral disapproval? (One could argue, for example, that despite widespread use, chemical pesticides are not a socially accepted technology).

- *Cognitive Parameters*: What's the level of awareness that a technology is being used? Are people presented with clear opportunities to accept or reject a technology? (Arguably, many of the technologies in daily use have not been "socially accepted" simply because they are, for the most part, wholly unknown.)

- *Technical Administrative Parameters:* Where is a given technology in the process of regulatory review and approval? Are there regulatory or court decisions that actively sanction the use of the technology?

- *Political Parameters:* What is the level of debate, contention and organized opposition to a technology? What is the potential to mobilize opposition at any given time, and at what cost?

There are also normative and quasi-normative parameters for the social acceptability of technology. Here are a few:

- *Religious Acceptability*: What do religious doctrines or religious authorities have to say about the technology? There are cases (such as whether cheese made with recombinant chymosin meets Kosher standards) where these are rather straightforward questions.

- *Cultural Acceptability*: Given the fact that many cultural norms are implicit and veiled, the question of whether a given technology is culturally acceptable is often vexed and contentious.

- *Ethical Acceptability*: I take this to be the broadest and most irreducibly normative category. *Should* society accept a given technology, and if so, under what constraints or qualifying conditions? (So, one might argue that a technology is acceptable only if its advantages and benefits will be distributed fairly across all economic classes).

Many of the social sciences employ forms of analysis that generate evaluations that are at least hypothetically ethical, in the above sense. Standard economic cost-benefit analysis is an obvious example. Though saying whether the costs of a technology outweigh its benefits is not a normative judgement in the sense of actually recommending for or against, such analyses nevertheless suggest qualified normative criteria that might be (and indeed commonly are) used for such a judgement. I hope it goes without saying that what "cost," "benefit" or "fairly" means is itself an ethical issue, that what it means for society to accept a technology (in a normative sense) is an ethical issue, and that whether there even *are* such normative criteria is an ethical issue. One can debate terminology, but the category of ethical acceptability is logically inescapable, since to deny the relevance or possibility of ethical criteria is to make a contestable normative claim.

Criteria for descriptive parameters for social acceptance tend to become entangled with normative criteria for social acceptability. The reasons have little to do with the study of technology *per se*. For example, behavioral social science may require framing assumptions about rationality in order to structure data bases or to characterize a behavioral phenomenon as an instance of "choice". But calling a form of judgment or behavior *irrational* is generally taken to imply a normative judgment about it. This opens out into both philosophical and highly tendentious debates, especially about the kinds of social science analysis noted above.

While it is tempting to dismiss talk about the social acceptability of technology and society's values as lacking proper rigor (Is society really the sort of thing that can have values?), it is useful to recognize some important things that are going on in such talk. Clearly, players — and we are all players here — try to "spin" the social acceptance of a technology such as nuclear power, food irradiation, e-commerce, or biotechnology in an effort to influence opinion. That is, social acceptance is a thoroughly reflexive phenomenon. Given the multiple ways (above) that a technology might or might not be characterized as socially accepted (or not), it is generally possible for those who wish to quiet or aggravate acceptance as measured by any given parameter to say (with some degree of truth) that the technology is/is not/may/may not be/will be/won't be socially accepted (with respect to some other parameter).

Given the economic, political and cultural and professional interests that are (or might potentially be) at stake, anyone may be an interested party. This is *not* to say that there are no facts with respect to the indices that might be taken to measure social acceptance of nanotechnology, nor to say that everyone wants to influence the social acceptance of a technology that they study. But it does suggest that any reported observation regarding these indices may with fairness be subjected to a normative critique. For my money, the reports of sources that disclose their interests will be more credible.

Most scientists (including social scientists) are not trained at clearly understanding their own interests, much less at disclosing them openly to others. Unfortunately, the arguments that are proffered in enduring philosophical debates, edifying in their own right and crucial to rigorous social science, are quite likely to be deployed in the strategic gamesmanship of players seeking to advance or retard the social acceptance of a given technology.

Has biotechnology been socially accepted? I will confine myself to food and agricultural biotechnology, which is my primary area of expertise. Certainly there are published reports that bear on each of the indices noted above.

- *Geographical Parameters*: One can measure the number of hectares sown in biotech crops, the countries in which they are grown, and the demographic characteristics of farmers who use biotech seeds.

- *Economic Parameters*: There are many economic studies of the adoption of biotech, as well as studies of the competitiveness of the industry.

- *Psycho-Social Parameters*: There have been repeated surveys of public opinion on biotech. Levels of acceptability vary in degree and in change of direction across national groups.

- *Affective Parameters*: Surveys and focus group research provide some indication of affective parameters. This is an ongoing area of data collection and qualitative analysis.

- *Cognitive Parameters*: Surveys indicate a general measure of awareness, generally quite low, it is worth noting, for a technology that has become a model case study for a contested technology.

- *Technical Administrative Parameters*: Since the early nineties there have been regular reports of regulatory approvals. The weakness in this area would appear to be the lack of a general theory of what might have been thought to be an obvious question: what constitutes legal or administrative acceptance?

- *Political Parameters*: A substantial amount of qualitative research exists. There do not appear to be standard methodologies for developing indices of political contentiousness and acceptability.

Some Lessons: Though a fair amount can be said about each of these parameters as they relate to food and agricultural biotechnology, what do we know about the social acceptance of food and agricultural biotechnology? I would be skeptical of anyone who professed to know whether this technology *has been* accepted or *will be* accepted, both on a global or a regional basis. While it seems unlikely that agricultural biotechnology will simply disappear, it is anyone's guess as to whether the levels measured in these indices are stable, or will increase or decrease over time.

To me, this suggests that we don't know very much about how these parameters affect one another, about the dynamics of social acceptance. The notion of social acceptance seems intuitively clear, and even demonstrable for some historical cases. For example, can we really question whether electrification has been socially accepted in the industrialized world? Yet I would question whether we really have a very good sense of what real-time social acceptance amounts to. The strategic, normative and reflexive dimensions of social acceptance may account for the open-ended nature of the real-time index for social acceptance. But it would be useful to have a theoretically clear and rich statement of why this is the case.

Has the battle of strategic positioning for the social acceptance of nanotechnology already begun? It is inevitable that, whatever the motives of their authors, the documents which already exist (the documents we are generating at this workshop) will be spun by players down the road. Because I have ideas about how which criteria and procedures should lead to the acceptance, rejection or qualification of any new technology, I would argue that several things should be done at the earliest opportunity. Most importantly, science funders should invest in the creation of fora that are shielded from strategic actors to the extent possible. There are, of course, limits to the extent that this *is* possible, but several of the following things would help. One is clear, open and ongoing multi-disciplinary clarification of the interests that are advanced and retarded by the development of nanotechnology. A second goal would be to increase scientists' capacity to reflectively understand the sense in which they are interested parties, and encourage disclosure of those interests. Here, career, disciplinary and ego-based interests can be as decisive as pecuniary ones. One should not shy away from development of explicitly normative studies and position papers, but such efforts should aspire to high standards of transparency, clarity of analysis and to the creation of a public record.

It would also be useful to build on the literature of social acceptance as it has been developed with respect to technologies such as biotechnology, to design studies that would compare and integrate the indices described above, and to examine analogies that might suggest a basis for normative evaluations of the social acceptability of nanotechnology. The "technology out of control theme" or the precautionary principle as applied to biotechnology might, for example, present a suggestive starting point for normative studies on the acceptability of nanotechnologies.

The scientists and engineers behind nanotechnology must be involved in all the above, and to me this suggests that there will be a need for an ongoing series of conferences, workshops, seminars and publications with a fairly high level of visibility. Such people will exclude themselves from the standard run of social science disciplinary research outlets, so something else is needed. These activities need to be visible enough so that someone who gets interested in the ethics or social acceptance of nanotechnology fifteen years from now won't need extraordinary good luck to find the public record. That's sorely missing with respect to agricultural biotechnology, where the social acceptance wheel is constantly being reinvented by players and naïve researchers alike. Perhaps that points to a national center, or perhaps to a program like the Ethical, Legal, and Social Implications of Human Genetics Research (ELSI) at the National Institutes of Health.

SOCIAL, ETHICAL, AND LEGAL IMPLICATIONS OF NANOTECHNOLOGY

Richard H. Smith, Coates & Jarratt, Inc.

The term "nanotechnology" may include such products as mono-layered materials, nanocomponents in smarter MEMS, and Fullerene-based computers. Some decades from now, it may also include communicating and/or programmable molecular machines. The implications discussed herein assume that some decades from now, the latter capabilities are achieved.

In the short-term, incremental improvements in processes and materials can come about by improved knowledge and skills in the nano-realm. The additive effects of these changes can have secondary and tertiary impacts that are transformational. For example, the social, ethical, and legal impacts of the World Wide Web were not originally thought to be considerable. The Web's enormous impacts are still not fully understood. In a similar way, nanotechnology may appear gradually and yet have a revolutionary effect. In the longer-term, the risks and rewards of nanosystems will certainly be exaggerated as our technological capabilities improve.

Virtually every "millennium survey" of the future poses some social, ethical, or legal questions about nanotechnology. But there is little so far in the way of serious study. Bill Joy's recent article, "Why the Future Doesn't Need Us" in *Wired* is a widely read example. Mr. Joy suggests that since nanotechnology is potentially dangerous, we should relinquish our study of it. His article, though intriguing, misses a most critical point. The United States doesn't have a monopoly on nanotechnology research. The rest of the world is spending over $1 billion per year in the field. Relinquishment by friendly governments, even in the unlikely event that it could be enforced, does not ensure that all researchers will make the same decision.

Others have also encouraged the public's fears about nanotechnology and biotechnology. Just as fear of cloning could slow efforts in biomedicine and fear of genetically modified foods could contribute to hunger, fear of futuristic nanobots running amok could delay the benefits offered by nanotechnology. When discourse is founded on emotional prejudices, its unreasonableness discredits the legitimate need to identify and assess risks. Nanotechnology could ultimately turn out to be risky, but the prudent way to assess the risks is not the abandonment of the field.

Asilomar demonstrated that the scientific community could design systems to contain high-risk technologies; the Shelter Island meetings showed us we could put great minds towards thinking about risk; the space program showed us we could have a dangerous program with almost no fatal missions. Notwithstanding the negativists, we have proven that we can manage the risks of powerful technologies. This does not suggest that we are safe — merely that we are not inevitably doomed to the worst of possible technological outcomes. We should hear from the pessimists, but we should not hear *only* from them.

Here are some examples of possible social, ethical, and legal implications that bear some thought and consideration. You have a handout that contains a more robust list.

Short Term (3-5 years)

- The NNI will fund careers for scientists and graduate students; commercial firms will invest in nanotechnology R&D. These efforts will be widely dispersed politically, geographically, and scientifically.

- Nanotechnology research will both require and produce enabling technologies that will have beneficial spin-offs.

- There will be unintended consequences, both good and bad.

- We will have to balance the opportunity costs of studying assembler-based nanotechnology (with its huge potential payoffs) against research in fields where advances might have nearer-term but smaller payoffs.

- Designs for future nanosystems will be created, challenged, modeled, improved.

Medium Term (5-15 years)

- Nanotechnology research may allow an otherwise moribund Moore's Law to continue operating.

- Fullerene-based computer chips may require enormously expensive fabrication facilities that turn out chips by the millions — needed for smart packaging, foods, Bluetooth devices, etc. This could result in a proliferation of inexpensive unit costs but prohibitively (and anti-competitively) expensive initial capital costs.

- We may discover and perfect nano-sized sensors and tools that can diagnose disease much sooner than ever before — perhaps long before we have cures.

- We should expect revolutionary advances in materials, MEMS, etc., resulting in abundant markets but also in an upheaval in global financial and manufacturing systems.

- Public education may be needed to balance the views expressed by anti-technology writers and press.

Long Term (over 20 years)

- Nanosystems may help solve problems of disease and aging, pollution and scarcity, overpopulation and starvation, and could create revolutionary changes unlike any ever seen.

- Nanosystems could help produce alternatives to fossil fuels and their high environmental price and reliance on foreign sources.

- The transition from a pre-nano to a post-nano world could be very traumatic and could exacerbate the problem of haves vs. have-nots. Have-nots do not easily obtain access to new technologies; the difference between the lives of the nano-rich and the nano-poor will likely be striking.

- Potential harmful uses — intentional and unintentional — need to be studied well in advance: nano weapons; intelligence-gathering devices; nanotechnology combined with Artificial Intelligence to form super-intelligent but virtually invisible devices; artificial viruses to which humans have no immunity, etc.

Recommended Social and Ethical Research Areas

- We should follow a broad path in the ethical studies of nanotechnology including utilitarian ethics (in its many forms), virtue ethics, communitarian ethics, deontological and religious studies, and views of administrative and distributive justice.

- We need to look at the broadest reasonable set of worldviews if we are to do justice to the problem.

- Scientists and engineers like those at Rice University, New York University, MITRE Corporation, and DARPA can inform us on what is feasible.

- Technology pessimists can probe areas of risk that might escape a less vigorous review.

- Optimists from organizations like the Foresight Institute can provide insights into the kinds of systems that seem to be within the realm of the possible if assemblers can be made to work.

- Organization and process professionals like those from ITRI and the Institute for Alternative Futures can coordinate experts who would typically not relate to each other.

- Technology assessment professionals like those with Coates & Jarratt, Inc. can apply time-tested tools to estimate a range of possible circumstances and assess the secondary and tertiary effects of incremental and radical changes.

- We need formal risk analysis because risk is one of the most important ethical and social issues we could imagine:
  - Risk of physical harm
  - Risk to economic and social systems
  - Risk to political and financial power bases
  - Risk of the haves fighting to keep the upper hand over the have-nots

- We need experts from several legal disciplines to understand the legal implications:
  - Litigators
  - Patent attorneys
  - Privacy law specialists
  - Constitutional lawyers
  - Academic legal scholars
  - Attorneys who understand the process of using the law as a weapon to retard progress

Most importantly, a multidisciplinary approach is critical to a satisfactory understanding of the social, ethical, and legal implications of nanotechnology. For this reason, there needs to be a gathering of all the PIs and other listed key personnel from each grant at least once every six to nine months with mandatory attendance and at least one paper presentation by each sponsored grantee institution (firm, individual, or university). This will not be inexpensive or easy to administer, but it will increase the odds of researchers understanding the issues.

At some point over the course of the next several years, we will have to think about how — scientifically — we are going to achieve a nanotechnology that includes communicating or computing capability. In order to assess the unintended consequences — secondary, tertiary, quaternary uses and effects of new technologies — we must consider what approaches are used. Otherwise, we can't possibly think forward to what unintended consequences might be or how they might be manifested. Nanotechnology arrived at through chemistry may have quite different characteristics than nanotechnology arrived at through biology.

On the following pages, there is a somewhat more comprehensive matrix of short-, mid- and long-term implications (tables 6.3, 6.4, and 6.5, respectively).

Table 6.3. Implications Matrix: Short-term

Time Frame	General characteristics
2003-2005	Mostly research, not completed projects or products. Some super-MEMS systems deployed. Nano-sensors being tested. Coatings and some other materials products ready or nearly ready. What we learn from the Human Genome Project and the study of proteomics adds more and more to the potential biological approaches to nanotechnology.

Table 6.3. Implications Matrix: Short-term (continued)

Social / Cultural Implications/Situations/ Questions	Ethical Implications/Situations/ Questions	Legal Implications/Situations/ Questions
The NNI will fund career steps for scientists and graduate students; commercial firms will invest in nanotechnology R&D. These efforts will be widely dispersed politically, geographically, technically and scientifically.	Should we try to modify nature (i.e., "play God")?	Who can patent what? What will be the general guidelines for the patentability of speculative capabilities like molecular modeling?

Table 6.3. Implications Matrix: Short-term (continued)

Social / Cultural Implications/Situations/ Questions	Ethical Implications/Situations/ Questions	Legal Implications/Situations/ Questions
Nanotechnology research will both require and produce enabling technologies that will have beneficial spin-offs.	Does nanotechnology offer any capabilities that will allow us to avoid animal testing in the future?	What governments have jurisdiction over research, patents, etc.? The states? The federal government? The EU, NAFTA, WTO? The United Nations?
We will come to understand that there will be unintended consequences, both good and bad.	Should we allow (or how can we avoid) political or religious interference in research funding? (E.g., an oil company might want to prevent research in a field that might eventually obviate the need for fossil fuels.)	Could global politics (e.g., the biotechnology fears in some European countries) slow nanotechnology research to the detriment of those who would go faster? Could social justice needs in LDNs accelerate development?
What might be the inadvertent effects of coatings, materials, etc.?	What ethical standards should be applied/considered? Deontological? Utilitarian? Communitarian? A combination? Who should decide who gets to decide?	Who will decide issues of government oversight vs. academic and industrial freedom?

Table 6.3. Implications Matrix: Short-term (continued)

Social / Cultural Implications/Situations/ Questions	Ethical Implications/Situations/ Questions	Legal Implications/Situations/ Questions
Designs for future nanosystems will be created, challenged, modeled, improved. How will the term "nanosystem" be applied? What will be allowed to count as a nanosystem?	Should the fruits of the research reach everyone or just the wealthy? Does this include only the U.S. or the world? If the world, who pays? Can global deployment be built-in, even for impoverished nations without substantial infrastructure?	Will any jurisdiction have veto power? What if Virginia wants to build a nano-experimental center. Could Maryland or DC veto it because of physical proximity? Could the Netherlands veto it because of potential danger? How could the risk be proven? Who would enforce the decisions?
What kinds of research get funding priority? Who pays: government or industry? Who owns the results?	What interest groups should get to debate the risks, costs, benefits, locales, etc. of nanotechnology research? Could any groups appropriately be excluded from the debate? Those who are uninformed (by whose standards)? Those without political power? Those who can't contribute to office-holders? How will the views and recommendations of those who are traditionally underrepresented be integrated?	Experts from several legal disciplines are needed to understand the legal implications: Litigators, patent attorneys, privacy law specialists, constitutional lawyers, academic legal scholars, attorneys who understand the process of using the law as a weapon to retard progress.

Table 6.3. Implications Matrix: Short-term (continued)

Social / Cultural Implications/Situations/ Questions	Ethical Implications/Situations/ Questions	Legal Implications/Situations/ Questions
Should we fund only those projects that have clearly definable scientific goals that are close to current capabilities or should we consider funding projects that seem more far afield from current capabilities?	We will have to balance the opportunity costs of studying assembler-based nanotechnology (with its huge potential payoffs) against research in fields where advances might have nearer-term but smaller payoffs.	
Should we relinquish the field to researchers from other countries? Researchers in the U.S. who don't follow the rules? Bad actors? How could we design and enforce a relinquishment policy anywhere?		
What kinds of risk assessment are possible when the technology in question is prospective? E.g., if nanotechnology results from molecular engineering, the secondary and tertiary uses (and risks) are different than if it is derived from modifications to DNA.		
How can we best encourage multi-disciplinary educational tracks like chemical physics and bioengineering?		

Table 6.3. Implications Matrix: Short-term (continued)

Social / Cultural Implications/Situations/ Questions	Ethical Implications/Situations/ Questions	Legal Implications/Situations/ Questions
How can we solve the problem of disciplinary bias in proposal writing? A logical means of approaching nanotechnology is through biology but biologists typically do not write proposals to the National Science Foundation. NIH is targeted for a relatively small proportion of the NNI funds. Do we create a task force to encourage/teach biologist NSF proposals? Do we create a new multidisciplinary funding entity? Should we consider outsourcing it to private industry?		

Table 6.4. Implications Matrix: Mid-term

Time Frame	General characteristics
2006-2015	Many super-MEMS products tested and in operation. Entirely new classes of materials and manufacturing processes are entering everyday life. Nano-realm diagnostic products are entering the consumer marketplace. Communicating and/or programmable nanosystems seem more and more plausible in the near future. Nanobots are on the observable horizon but not testable yet.

Table 6.4. Implications Matrix: Mid-term (continued)

Social / Cultural Implications/Situations/ Questions	Ethical Implications/Situations/ Questions	Legal Implications/Situations/ Questions
Nanotechnology research may keep Moore's Law operating when it otherwise might hit a wall.	We may discover and perfect nano-sized sensors and tools that can diagnose disease much sooner than ever before — perhaps long before we have cures.	We should expect revolutionary advances in materials, MEMS, etc., resulting in abundant markets but also in an upheaval in global financial and manufacturing systems.
Fullerene-based computer chips may require enormously expensive fabrication facilities that turn out chips by the millions — needed for smart packaging, foods, Bluetooth devices, etc. This could result in a proliferation of inexpensive unit costs but prohibitively (and anti-competitively) expensive initial capital costs.	Who might be marginalized by the realization of nano-capabilities? The poor (as usual) or the currently powerful?	If we ever approach what Ray Kurzweil calls "The Age of Spiritual Machines" (i.e., computers that claim to have human capabilities) what precedents will be applied? Will new laws be needed?
What role should educators play in balancing the views expressed by anti-technology writers and press.	What are the implications of sensors on privacy? MEMS-based sensors could be anywhere and not seen. DNA sensors might be even more invasive. Yet each could have valuable properties (e.g., medical, investigative, etc.)	Who will decide questions about "snooping" by MEMS-based sensors? The same groups that decide on sensors bugs?
Who will control the means of production? The government until safety is assured? The industrial or academic investigator that produces the original product or process?	Who should benefit from the longer-term discoveries? Only those who can pay for them? Everyone? Only those in the country of discovery?	Will MEMS-based devices cause harm to patients? If so, what standards will be used to ascertain medical malpractice?

Table 6.4. Implications Matrix: Mid-term (continued)

Social / Cultural Implications/Situations/ Questions	Ethical Implications/Situations/ Questions	Legal Implications/Situations/ Questions
As sensing goes from parts-per-million to parts-per-billion to parts-per-trillion, etc., what are the implications for environmentalism, crime fighting, etc.?	What are the questions we should ask about clinical trials of nano-devices, nano-diagnostics, etc.? Does this question imply that nanomedicine should be restricted somehow? To academic medical centers? To some other regulated institution? Should the market decide?	Will there be limits on the liability of researchers and their employers? Who will cover the cost?
Who would best conduct comprehensive risk assessment? If developers and researchers are wrong in their assessment of risks, who might suffer and how much?	What is the most appropriate site of experiment in early stage nano-device research? White paper only? Computer model? Physical model? Animal? Human? Who will decide these questions?	Is a nanomedical treatment a drug or a device? What and whose rules apply?
What are the impacts on the Western democracies if the first substantive discoveries are made by an unfriendly or undemocratic government or non-state actor?	How, when, and by whom will standards be set for tracking devices, tagants, etc.?	What will/should be the roles of the Food and Drug Administration? The Environmental Protection Agency? The Customs Service?
What industries will become marginalized? Will governments have to subsidize newly irrelevant businesses?	Might an increasingly sophisticated nanomedicine capability change the way society looks at "risky" behaviors? Is this good or bad?	

Table 6.4. Implications Matrix: Mid-term (continued)

Social / Cultural Implications/Situations/ Questions	Ethical Implications/Situations/ Questions	Legal Implications/Situations/ Questions
Will new interest group coalitions be established while others recede? For example, will potential environmental clean-up capabilities offset higher risk mineral exploration?		

Table 6.5. Implications Matrix: Long-term

Time Frame	General characteristics
2020	Communicating and/or programmable nanosystems are becoming available. Nanobots are working in labs and being tested, evaluated, and fielded for various specific applications. Nanomedicine is replacing older forms of medicine such as surgery, traditional pharmaceuticals, rational drug design, etc. Universal assemblers are still not available.

Table 6.5. Implications Matrix: Long-term (continued)

Social / Cultural Implications/Situations/ Questions	Ethical Implications/Situations/ Questions	Legal Implications/Situations/ Questions
Nanosystems may help solve problems of disease and aging, pollution and scarcity, overpopulation and starvation. Assembler-based systems, if they are found to be feasible, could create revolutionary changes unlike any ever seen.	Potential harmful uses — intentional and unintentional — need to be studied well in advance: nano weapons; intelligence-gathering devices; nanotechnology combined with Artificial Intelligence to form super-intelligent but virtually invisible devices; artificial viruses to which humans have no immunity, etc.	Who will represent the rights of patients of nanomedical procedures? What will constitute "informed consent"?
Nanosystems could help produce alternatives to fossil fuels and their high environmental price and reliance on foreign sources.	The transition from a pre-nano to a post-nano world could be very traumatic and could exacerbate the problem of haves vs. have-nots. Have-nots do not easily obtain access to new technologies; the difference between the lives of the nano-rich and the nano-poor will likely be striking.	Who should / could regulate nano-weapons? How will verification be accomplished?
Most socio-economic systems are based on scarcity. What will happen if nanotechnology allows scarcity to become scarce? What would happen to the concepts of wealth? Power?	How much nano-prosthesis will make one non-human?	How would the concept of property change if most things became replicable? Will we care?

Table 6.5. Implications Matrix: Long-term (continued)

Social / Cultural Implications/Situations/ Questions	Ethical Implications/Situations/ Questions	Legal Implications/Situations/ Questions
If nanotechnology is as transformative as some of the nano-optimists think, how tough will the transformation be? Who should be in charge of softening the blow?	Can or should we consider the replication of brains? Souls?	If assemblers or communicating / computing nanobots are ever constructed, who would own them? The manufacturers? The public? Individual citizens?
If nanotechnology changes the manufacturing enterprise so much that common laborers are no longer needed, what would be the effects on employment, family structures, recreation, leisure time, the arts?	What might be the implications of truly sentient artificial intelligences? Are we forced to view these potential entities as foreign and destructive? Is any "intelligent" machine necessarily a terminator?	Will we need the nano equivalent of a "Terminator seed" for nanobots? Engineered apoptosis?
Are there any implications for the space program that bear the attention of long-range planners? Might we be able to reduce payload to energy ratios sufficiently to allow new programs? Could we use other planetary bodies as quarantined nanotechnology test-beds? Might we consider terraforming other planets?	Will the nature of man change? Is this good? Is this bad? Who gets to decide?	
If the developed world generates the technological changes, will it have the right to impose the degree and pace of change on the rest of the world?	Will the "Haves" of the Earth no longer need the "Have-nots"?	

Table 6.5. Implications Matrix: Long-term (continued)

Social / Cultural Implications/Situations/ Questions	Ethical Implications/Situations/ Questions	Legal Implications/Situations/ Questions
How will/should humans interact with nanobots?	If nanotechnology is as transformational as some suggest, will people have the right (or a way) to opt out of such a society?	

ENVISIONING LIFE ON THE NANO-FRONTIER

Mark C. Suchman, Sociology and Law, University of Wisconsin – Madison

In the spirit of Bonnie Nardi's paper in this volume (Nardi 2001), I would like to "envision" some of the legal and organizational challenges that might arise from the widespread introduction of nanotechnology into the U.S. economy. Obviously, true prediction currently lies beyond the capabilities of even the most informed and foresightful experts in the field, let alone an outside observer like myself. Nonetheless, social scientific research on previous technological revolutions suggests at least a few tentative hypotheses. In particular, as we speculate on policy implications, we might be well advised to distinguish between two types of nanotechnology — which I will call nano-materials and nano-mechanisms, respectively — and to explore the implications of each separately. I suspect that the social disruptions and governance challenges stemming from the former are likely to be much more localized and manageable than those stemming from the latter.

Two Flavors of Nanotechnology

Accounts of the coming nanotechnology revolution seem to involve two somewhat separable agendas — each with its own distinctive social implications. The first (and, in the short run, probably more significant) agenda links nanotechnology to chemical engineering and material science. This variety of nanotechnology focuses on controlling the nano-scale organization of macro-scale substances. Examples of such "nano-materials" — or "nanates" — might include wear-resistant nanostructured polymers for tires and drive belts, super-hard nanostructured ceramics for drill bits and cutting surfaces, or ultra-fine nanostructured membranes for filters and seals.

The second (and, in the long run, probably more disruptive) agenda links nanotechnology to mechanical engineering and robotics. This variety focuses on constructing nano-scale devices for operation in macro-scale environments. Examples of such "nano-mechanisms" — or "nanites" — might include ultra-small in-vivo medical devices, miniaturized surveillance systems, or lilliputian mining and manufacturing equipment.

Needless to say, these two varieties of nanotechnology blur together at their boundaries: One could, for example, imagine a nano-material filter that employed interconnected nano-mechanism "turnstiles"; or, conversely, one could imagine a nano-mechanism robot that assembled, inspected or repaired nano-material compounds. However, the broad distinction between nano-materials and nano-mechanisms considered as archetypes serves to highlight important "scope conditions" for predictions about how nanotechnology will (or will not) transform the social order.

Nano-Materials as Discrete Technological Discontinuities

For many industries and many aspects of social life, nanostructured materials are likely to represent profoundly important technological developments. Indeed, if nanates live up to their early billing, the resulting societal transformations could equal the transformations that attended the development of bronze implements at the end of the stone age or the introduction of nuclear weapons in the 20th century. Given this, it may seem odd to predict that the impacts of nano-materials will be "localized and manageable," as asserted above.

My premise, however, is that nano-materials with enhanced performance characteristics do not, in and of themselves, pose unprecedented challenges to social organization. Admittedly, particular new materials may engender the development of particular new products that profoundly transform particular areas of social life — and the ripples from such transformations may spread through society in complex and unpredictable ways. (One need but imagine the implications of a bullet coating that allowed small arms fire to penetrate tank armor, or of a photovoltaic cell that eliminated the need for fossil fuels.) But such transformative potential is hardly unique to nanoscience. Nanotechnology may be more likely to yield these radically new materials, but humanity has developed quite a few other transformative compounds — from glass to gasoline to plastic — without the aid of nano-scale understanding. Thus, although the introduction of new nanates may be revolutionary, it promises to be revolutionary in relatively familiar ways.

Recent years have generated a substantial literature on the industrial impact of technological discontinuities (for a survey, see Tushman and Anderson 1996). In brief, the central finding of this literature is that the introduction of any radically new technology initiates a period of ferment within the affected industrial sectors. During this ferment, businesses, public agencies and other "technology champions" jockey for position, attempting to mobilize various economic, cultural and political forces to frame and tame the discontinuity, and to thereby establish a new "dominant design." Once such a dominant design emerges, however, the affected sectors restabilize — although often in significantly different configurations than before. Presumably, a similar dynamic would attend the introduction of new nano-materials, albeit perhaps at a higher intensity and a broader scale.

Of course, this hardly obviates the need for foresightful policy attention: The changes initiated by even a relatively modest technological discontinuity can be quite far-reaching, and the framing and taming process can be intensely political, both within the affected industries themselves and at the level of the larger polity. However, the differences between previous technological discontinuities and the discontinuities resulting from nano-materials seem likely to be matters of degree, rather than of kind. Industries will certainly be transformed, and some of these transformations will be wrenching and risky, but the involvement of nanotechnology will not, in itself, make the transformations any more wrenching and risky than the transformations that we have seen in the past. It is in this sense that the impact of nano-materials will be "localized and manageable": The policy issues, even if large, will arise from the particular performance characteristics of particular products, not from the inherent nature of nanotechnology per se. If so, case-by-case planning would appear to represent an appropriate and, arguably, sufficient response.

(As an aside, it may be worth noting that nano-materials could produce unprecedented disruptions if they instigated technological discontinuities in an unusually large number of industries simultaneously. Without entirely ruling out such simultaneity, however, the practical limits on human resources and attention strongly mitigate against this prospect. In all likelihood, applications of nanotechnology will be neither more nor less staggered in their arrival than applications of previous multi-purpose technologies, like semiconductors, synthetic polymers, or wireless telecommunications.)

Nano-mechanisms as Unique Governance Challenges

In contrast to the relatively familiar challenges of nano-materials, nano-mechanisms promise (or threaten) to confront society with policy issues that are as unprecedented as they are profound. By allowing humans to manipulate the world at a previously unattainable scale, nano-mechanisms open a genuinely new frontier, beyond the contemplation of traditional legal and governmental regimes. While, as Richard Feynman put it, "there is plenty of room at the bottom," (Feynman 1961) at the moment there are very few sheriffs at the bottom, to keep that room safe and productive.

As currently envisioned, nano-mechanisms are likely to possess at least three properties that will generate novel safety and governance challenges: invisibility, micro-locomotion, and self-replication.

1. Invisibility is, of course, an inherent property of nano-scale objects, whether natural or artificial. Artificial nanites, however, would be among the first complex constructions intentionally engineered to accomplish human purposes at a microscopic (or sub-microscopic) level, and their introduction into the technological armamentarium would dramatically increase the potential for orchestrated covert activities.

2. Micro-locomotion is less inherent in nanotechnology than is invisibility, since some nano-mechanisms will undoubtedly work best when anchored to substrates. Nonetheless, free ranging nanites have clear advantages for certain tasks, and as such devices are introduced, they will radically challenge traditional understandings of macro-boundaries and barriers. Fences, walls and even human skin are largely open space, at the nano-scale.

3. Self-replication is clearly not an inherent property of nano-mechanisms, and indeed, creating self-replicating nanites may prove to be one of the most difficult technical hurdles of the nanotechnology revolution. Self-replication, however, is essential for the economical production of complex nano-mechanisms in useful quantities, and thus it seems likely that by the time such nano-mechanisms become socially significant, self-replication will in fact be a fairly common attribute. Unfortunately, self-replication also poses profound challenges to human foresight and control, since a population of carelessly designed self-replicating nanites could grow exponentially, without a ready "off switch."

Needless to say, the hazards of invisibility, micro-locomotion and self-replication would be magnified if nanites also possessed a capacity for autonomous operation and self-modification; however, such higher-order artificial intelligence is not a prerequisite for the envisioned crisis of nano-governance. Even "dumb" nano-mechanisms would require a dramatic rethinking of society's current legal and normative structures.

Three governance issues promise to be particularly acute: monitoring, ownership, and control. Moreover, these issues may play out not merely with respect to nano-mechanisms themselves, but also with respect to the "nanospace" that such mechanisms render accessible:

1. Monitoring: Invisible, micro-locomoting, self-replicating nanites will severely test society's established assumptions about what can and should be monitored, and by whom. On the one hand, the lay public may find itself inundated (or even interpenetrated) by complex devices whose presence, provenance and purpose remain undetectable without sophisticated technical assistance. On the other hand, those actors that can produce or purchase nanites will find themselves able to monitor their worlds (including their social worlds) in more profound and surreptitious ways than ever before. Contrary to traditional assumptions, ordinary people will no longer be able to observe all the socially relevant activities in their own surroundings; yet public authorities will be hard pressed to provide policing assistance without further endangering individual privacy. In the absence of trustworthy institutions to regulate this newly-opened invisible frontier, the potential for rampant abuse will be matched only by the potential for rampant paranoia.

2. Ownership: Nano-mechanisms will also severely test society's assumptions about what can and should be owned, and by whom. Self-replication raises difficult questions of whether property rights persist from one generation of mechanisms to the next — not to mention difficult practical problems of "branding" proprietary nanites and policing "nanite rustling." Equally importantly, micro-locomotion raises difficult questions of who owns the nanospace through which nanites pass. Controversies akin to historical debates over airspace and rights-of-way may emerge over the nano-spaces within macroscopically "solid" objects. If your nanites get sucked into my air conditioning system, are you trespassing or am I kidnapping? Can public health authorities claim an easement for the passage of biomedical nano-sensors through my gut? If nano-robots in my tap water "harvest" a microscopic quantity of copper from my plumbing, is their owner guilty of breaking and entering?

3. Finally, and perhaps most significantly, nano-mechanisms will test established assumptions about responsibility and control. Traditionally, legal liability for mechanical devices can outlast ownership, but can be

attenuated by unforeseeable circumstances or by the intervention of a third party's will. The viability of these principles, however, becomes unclear in a world of self-replicating nano-mechanisms that operate invisibly in an only partially understood micro-environment. One might imagine governing the control of nano-mechanisms by analogy to defective products, or to toxic emissions, or to speeding vehicles, or to straying livestock — but the implications of these analogies are not entirely congruent with one another. Moreover, a parallel set of complexities and contradictions surround the control of nanospace: Who is responsible for policing the boundaries between "open range" and "enclosed territory" at the nanoscale? Can negligently maintained nanospace pose an "attractive nuisance" to passing nanites? Does a state maintain territorial rights to the nanospace within the bodies of citizens traveling abroad?

Conclusion

Although the distinction between nano-materials and nano-mechanisms is to some extent an artificial one, it highlights an important dimension of differentiation in the impact of nanotechnology: Nanostructured materials may pose serious practical and ethical challenges for particular policy domains, but these challenges will arise at a familiar macro scale, for which we have numerous rules, institutions and historical precedents. In contrast, nano-engineered mechanisms will force us to reformulate our rules and institutions to govern an unfamiliar setting with which we have no prior experience. By allowing human will to operate in the previously unreachable interstices of macro-reality, the nanotechnology revolution promises to open vast new frontiers for exploration, exploitation and settlement. Taming those frontiers, however, will require us not only to control physical matter, but also to control ourselves.

References

Feynman, R.P. 1961. "There is plenty of room at the bottom." In *Miniaturization*. New York: Reinhold.

Nardi, B. 2001. "Cultural Ecology of Nanotechnology" in this volume: *Societal Implications of Nanoscience and Nanotechnology*. National Science Foundation, Arlington, VA.

Tushman, Michael and Philip Anderson (1996), Managing Strategic Innovation and Change: A Collection of Readings. London: Oxford University Press.

SOCIETAL IMPLICATIONS OF NANOTECHNOLOGY

M.J. Heller, Nanogen

In the past my thoughts and ideas about nanotechnology have flowed easily, as they were prompted by the excitement and prospects of a coming new wave of scientific challenges and opportunities. Putting together a statement on the societal implications of nanotechnology has proven more difficult. During the past six months I have asked a number of people what they thought nanotechnology meant, and what they thought the societal implications might be. While most responses were positive, they varied depending on the degree of technical background. Most scientists and engineers saw nanotechnology as an encompassing technology that would broadly impact on microelectronics, data storage and the computer area, energy conversion and chemical processing, defense and national security, biotechnology (including agriculture), and the biomedical areas of therapeutics and diagnostics. While some scientists and engineers felt that nanotechnology was not quite as "new" as perceived and presented by the popular media, they did agree that a new level of much higher activity and broader applications was definitely happening. For the most part, this group felt that the overall implications of nanotechnology would be very good. In fact, some felt it might even be vital that we progress as rapidly as possible into this new area. The feeling was that nanotechnology (as well as other related technologies) may provide solutions to some of the more intractable problems related to eliminating disease and famine in developing countries, and helping to improve the economy and general living standard for the rest of the world. In general, scientists and engineers that I talked with tended to feel nanotechnology would carry risks similar to those seen in the development of other previous technology revolutions, e.g., microelectronics. The feeling was that, while nanotechnology could certainly be used for some insidious purposes, the overall benefits would far outweigh the risks. If there was concern about nanotechnology, it centered more on the need to be careful not to oversell or make inflated promises about nanotechnology, as this could lead to loss of credibility and confidence by the general public.

Responses from non-technical professional people were different from those of the scientists and engineers. In general, this group was not as well aware of the potentially more far-reaching applications of nanotechnology (and this would not be unexpected). This group tended to equate nanotechnology primarily with the further miniaturization of microelectronic devices and very advanced computers. Some from the non-technical group related to nano-type medical devices, which might for example be put into the blood

277

stream of humans to treat cancer or other diseases. While this group may not really understand nanotechnology, they generally thought that it was both very interesting and important, and that it might provide many future benefits. Some from both non-technical and technical groups do have an underlying uneasiness with how quickly technology in general is advancing. This uneasiness could most certainly reflect on nanotechnology. Some concerns about nanotechnology (and technology in general) are not just related to deliberate or insidious misuse of these technologies, but also include problems which could result from carelessness or unexpected adverse effects which appear later in implementing the technology. By way of previous examples, the dangers from nuclear power plant breakdowns, the pollution from chemical processing and toxic reactions to certain therapeutic agents. While these types of concerns are justified, most people still believe that technology will of course continue to advance in the future and will most certainly provide benefits for all of mankind.

SOCIO-ECONOMIC RESEARCH ON NANOSCALE SCIENCE AND TECHNOLOGY: A EUROPEAN OVERVIEW AND ILLUSTRATION

M. Meyer, Technopolis Ltd.

Summary

The study of societal implications of nanoscale science and technology (NST) is an area where science meets social science. Such interfaces often provide some potential for confusion. Scientists and technologists on the one hand and social scientists on the other do not share a common language. They use the jargons that are typical for their area of activity.

This is true even more so in a field where two different systems of science come together. Natural scientists and technologists have to understand elementary notions of social science, and social scientists in turn have to learn about the scientific and technological object of their study to do a meaningful analysis. The basic vocabularies have to be learned yet.

The objective of this paper is to provide a European point of view on the social implications of NST. Past and ongoing efforts in the social sciences that are related to NST are introduced. Part I of this paper will give an overview of mainly European studies in this area. Over the past few years, a modest amount of activity of social scientific research has been carried out in Europe that specifically addressed economic and social issues of nanotechnology. A review of these activities might provide a useful platform for further discussions.

Part II will give an illustration of a quantitative social scientific approach to NST. This is done to clarify where social sciences can contribute and where inputs from actors in the field are needed. Finally a number of conclusions are drawn and some suggestions are made with respect to where future social scientific research on NST might be of interest.

Part I: An Overview of Ongoing Socio-Economic Research

Economic and Social Aspects Of Nanotechnology

One should make a distinction between activities in the social science arena that are directed at describing the emergence of nanoscience and technology as it happens and those activities that deal with issues that might become relevant once nanotechnology has entered a more mature stage in its development.

Studies on Current NST Activities or Near-Term applications

Often and especially in a European context, nanotechnology or nanoscience is discussed in terms of potential paths towards commercialization, niche markets for already or soon-to-be available products. This is also the major focus of ESANT, the European working group for Economic and Social Aspects of Nanotechnology, a small community of researchers and practitioners, mostly with a technological background. The group has published a report (http://www.nano.org.uk/ESANT99.htm). In their report, potential avenues of near-term commercial development of nanotechnology are outlined.

Other social science research focuses on how governments support nano-scientific research and the implications this has for the development of nanotechnology. One example is the work commissioned by the Finnish National Technology Agency (Meyer 2000b). Table 6.6 gives an overview of the policy-relevant findings of the study. Nanotechnology is addressed in many different ways. The different perspectives observed at the country-level seem to coincide with the strengths of the respective national innovation system. There is no common definition as to what nanotechnology comprises. It appears that specific scientific and technological communities have reached a common basis as to what "nano" means with respect to their particular area. The current state of development allows government actors to play a substantial role in developing a novel technological area.

In some European countries, governments rely to a considerable extent on technology foresight studies in their long-term science and technology policy planning. These exercises provide governments with estimates about future technological developments and the implications they might have for society. Some of these studies also addressed NST. We shall have a closer look at them in the following section.

Table 6.6. Research Findings in a Policy Context

	Countries	Technologies	Industries	Firms
How is new knowledge explored, accessed and integrated?	National emphases seem to vary with the different overall strengths and weaknesses of the various national systems of innovations: Nano-electronics in the U.S., nano-materials and their auto and chemical applications in Germany, biomedical and instrumentation in the UK, materials in relation to ITC and pharmaceuticals in Sweden, nanophysics focus in Finland.	Technology-specific understanding of what nanotechnology comprises: Typical of an emerging field, there are a number of competing or complementary definitions. Disciplinary and sectorial boundaries are not overcome. Frequently "nano" is said to start at the 100 nm level, but in electronics the term is often associated with effects at the < 30 nm level; while in the materials sciences and technologies a line is drawn at ~ 300 nm. Some technologists in this area use "nano" as a new label for their sub-micron (< 1,000 nm) activities.	Firms address nanotechnology issues in a manner that is typical of their particular sector: Instrumentation, tools & techniques: networks of predominantly small firms, integrated R&D activities, innovations user-stimulated; chemicals: research-intensive sector, strong industrial in-house research, substantial number of patents, manifold contacts with academic researchers, large firms, but also a number of smaller nano-materials suppliers;	Chemicals is a sector that uses as well as supplies (intermediary) products. Pharmaceuticals: There are considerable contacts with nanotool-makers. Since nanotechnology can be viewed as an instrument-driven technology, nano-resolution instruments were the first to be commercialized. Not surprisingly, one can find a number of university spin-offs in this area. However, it is subject to debate if nano-tools are an aspect of nanotechnology or not.

Table 6.6. Research Findings in a Policy Context (continued)

	Countries	Technologies	Industries	Firms
		Not only differences, but also considerable overlaps between industrial sectors and their interest in nanotechnology around <u>technological themes:</u> Nano-resolution instrumentation used in different sectors and possibilities for their applications are a theme that "unites" the dispersed nano-communities. In a similar manner, ultra-thin films are viewed as important to different sectors, which is indicated by developments of industrial research activities. Electronics firms move into chemistry-related research, and vice versa. In a similar manner, nanostructured materials of interest to different user industries. This suggests clustering of activities according to technological themes.	<u>pharmaceuticals:</u> large firms, orientation dependent on affiliation with a chemicals "mother" or not, biomedical applications of importance, including drug delivery systems, established links with academia, but large firms set up close-to-academe groups to access the novel, until recently lacking knowledge; <u>automobile:</u> large manufacturers interested in potential nano-applications, research in nano-materials and nano-electronics closely linked to suppliers; <u>electronics:</u> large firms, R&D focused on semiconductors and lithography at dimensions smaller than 50 nm, first applications: sensors and actuators; <u>telecoms:</u> companies see no direct applications currently, but room is given to individual employees pursuing	

Table 6.6. Research Findings in a Policy Context (continued)

	Countries	Technologies	Industries	Firms
			activities. Large telecom companies that have a basic research base pursue nanoscale research related to electronics and chemistry	
What roles have political measures played so far?	Variety of approaches (see below). National policy measures so far were focused on awareness creation. Some efforts aimed at network facilitation, university-industry, interdisciplinary, and inter-industry collaboration: U.S.: Special programs at the level of at least 12 government agencies, loosely coordinated by an informal inter-agency working group; UK: No current dedicated nano-activities, university-industry LINK program until mid-1990s, including networking activities, managed program at a RC; Germany: funding of individual projects and	Two types of programs: S&T programs: [i] explicitly nano-related programs [ii] programs related to other S&T fields Programs serving other purposes (e.g., business development, human capital development, university-industry collaboration, local development, and incubator & networking programs). Rarely explicitly addressing nanotechnology in terms of technological clusters. Technological themes sometimes addressed in non-"nano" programs (e.g. specific aspects of materials research where also nano-	Policy support tends to address specific technological topics. Projects funded are often of an entirely academic nature. Where industrial partners are involved, projects might be narrowly defined so that just one partner from a user industry might be involved. At the program level, the programs are very inclusive. Usually, a variety of user industries are represented. More research-intensive industries have some exploration activities installed with their industrial associations. A program, such as the UK LINK Program, which demanded the participation of an industrial partner, leads automatically to a very selective focus on one or	ATP seems to be the most appreciated policy tool of successful nano-start ups. Less commercial nano-ventures in the U.S. appear to be associated with research grants from institutions like NSF and DOE. Occasionally, spin-offs identified these moneys as major source of income (together with "love money" of the founders) even several years after their establishment. More successful cases of start-up companies are characterized by integration in extensive business networks.

Table 6.6. Research Findings in a Policy Context (continued)

	Countries	Technologies	Industries	Firms
	project clusters, and workshops, currently (until 2003) support for the set-up of "Centers of Competence in Nanotechnology"; Sweden: nm-activities in university-industry "interdisciplinary materials consortia" over the past 10 years, also a nano-chemistry program has been set up; Finland: National Nanotechnology Program (Tekes/SA), project-based research funding.	aspects are important [ii]). Some nano-activity was located in biotech-based business development clusters.	two industries only.	

Mid- and Long-Term Issues

Foresight studies attempt to depict an image of a possible future using a variety of techniques. An important foresight method is the Delphi survey. A Delphi survey basically is a tool to create consensus and detect areas of conflicting expert opinions. In a first round, experts are confronted with a number of topics they have to evaluate with respect to time of realization, implication on wealth creation, quality of life, and similar issues. In another round the results of the previous round are introduced to the experts who then have a chance to re-evaluate the topic. Table 6.7 gives an overview of the nanotechnology-related topics covered in the 1995 German Mini-Delphi survey.

Kuusi and Meyer try to distill five major areas of developments ("*leitbilds*") drawing on this Delphi exercise and information from other technology studies (Kuusi 1999). *Leitbild* is a loan word from German, meaning a guiding image. Within the context of the social studies of science and technology, it can denote developments in unfolding areas that might lead to the emergence of a novel paradigm or technological trajectory. Its major function is to provide a platform integrating actors from different areas by providing a common goal, which can be an envisaged technique or product.

Table 6.7. Nanotechnology Topics in the German 1995 Mini-Delphi Study

Section "Cognitive Systems, Artificial Intelligence & Nanotechnology, Microsystems Technology", Subsection "Nanotechnology"	Period of Realization
14 – Functional materials and/or semiconductor components whose compositions and dotting densities vary from atomic layer to layer are widely used.	2006-2010
15 – Electronic solid-state components that consist of "super atoms" of artificially composed atoms will be developed.	2006-2010
16 – Methods to synthesize substances with new functions (e.g., polymer crystals with weak bonds) will be developed by way of combining various types of bonds at the atomic level.	2006-2010
17 – Nanostructured materials with predetermined properties will be manufactured.	2001-2005
18 – Organic hybrid composite materials that are based on the control of mono-molecular layers will be developed.	2006-2010
19 – Organic-inorganic composite materials will be developed (e.g., biomimetically) whose elements are at the level between several and a few dozen nanometers.	2001-2010
20 – An analytical method that sorts out a particular type of atoms using high-definition surface-analysis techniques will be in practical use.	2001-2005
21 – "Atomic function elements" (atomic switches, atom relay transistor, etc; in which movements of a small number of atoms cause logical and/or storage functions) will be in practical use and have a higher reliability and processing velocity than solid-state components.	2011-2015
22 – Reaction and synthesis methods at individual atoms or molecules of respectively atomic or molecular level of magnitude will be in use applying techniques from scanning tunneling microscopy.	2006-2010
23 – Techniques to fabricate structures at the atomic level that will not be based on probing methods as represented by the scanning probe method will be in practical use.	2006-2010
B – Organic, molecular composed materials will be developed using the natural method of self-organization.	2006-2010

Source: BMBF 1996.

1. Nano-resolution methods of analysis. One *leitbild* is that of nano-resolution analytical methods (topics 20, 22). The aim here is to further improve existing tools by adding new functions to analysis tools. Here,

realized techniques are further generalized into promising tools. The development of analytical tools and techniques is closely related to progress in other areas.

Often nano-resolution tools are identified with nano-resolution optical microscopy since these methods combine the possibility of measurement with that of manipulation, which makes them more versatile as tools for nanotechnology. A number of scanning probe microscopes have been developed and some of these methods do not require complicated sample preparation; e.g., the ability to work on *in vivo* substrates and determine structure-function relationships is the main reason why the AFM, the atomic force microscope, is so popular amongst biologists. It is said that even (nearly) two decades after its invention, the implication of the scanning-tunneling microscope (STM) and its follow-ups is still growing. In conjunction with continuous technical further development of this method, researchers are discovering new phenomena in the fields of physics, chemistry, and biology. At the same time the STM methods are used as a nano-tool rather than a nano-probe. The idea is to modify surfaces and tailor their structures on the nanoscale, down to the manipulation of individual atoms (Frenken 1998).

This case has shown that realized techniques — such as AFM and STM — can provide a basis for developing more sophisticated "promising" techniques that will go beyond mere measurement. Ultimately, they might facilitate large-scale manipulation at the nanometer level.

2. Nano-materials.   Another *leitbild* ranges around the notion of nanostructured materials. Nano-materials can be seen as structures that have particular properties owing to their size. The *leitbild* in this context is to manufacture a nano-material in a way that allows predetermination of its properties. This generalization pattern describes the transition from realized techniques to promising targets. Paired with a better scientific understanding of the subject matter, a variety of already realized techniques allow us to develop rather concrete ideas of improved materials. The idea here is to take advantage of nanoscale characteristics of structures and substances to create new materials with enhanced properties, such as those common to polymers, composites, or other materials (topics 16, 17, 19). The aim is not necessarily direct control of individual atoms; bulk operations suffice to exploit the nanoscale properties.

One example to illustrate the idea of bulk-processing nano-materials is that of colloidal dispersions (Philipse 1998). Colloidal science deals with the physics and chemistry of finely dispersed matter. Colloids are generally understood as particles or other objects with at least one dimension roughly in the sub-micron range. Nano-particles then are viewed as colloids smaller than 100 nanometer. It is pointed out that colloid science has a long tradition involving nano-particles and that not all that is nano is necessarily new (Philipse 1998). In this sense, colloids encompass gold colloids, colloidal silica, and aluminum oxide powders. Due to their small dimensions, colloids exhibit Brownian Motion. Owing to their large surface area, surface forces and repulsions determine the interaction between colloidal particles in the liquid phase. The balance between these forces depends critically on the details of the particle surface and the liquid composition. Colloids aggregate easily to form large flocs, networks or gels. The control and understanding of these aggregation processes is the major challenge of colloidal nanoscience. Suspensions of colloids, or dispersions, are, for example, milk, blood, cosmetics, such as toothpaste, or ink. Examples for inorganic colloids are clay particles, iron oxides on computer disks and in magnetic fluids, pigments in paints, or powders for technical ceramics. Colloidal systems are studied from a (bulk) chemical synthesis perspective. Such research integrates the study of physical properties of dispersions and gelation phenomena, obtaining important input from computer simulations and statistical mechanics. It is said that this wide scope of colloidal science is unavoidable, since finely dispersed matter can be encountered in many disciplines and applications.

The nano-materials *leitbild* provides a focus point for integrating this and other bulk chemistry topics with related activities.

3. Ultra-thin layers. A number of topics also revolve around thin layers (topics 14, 18). Here, efforts appear to be directed at characterizing these structures.

Thin-film technologies constitute a considerably well developed field. An exact ultra-fine production of thin films is a necessary condition for the subsequent characterization. Designing ultra-thin layers can be associated with a small number of technical aims, such as atomically exact delineations of layers, quantumized potential distribution, defined pore distribution in layers, ultra-thin separation and protection layers, improved layer function by way of multilayer structuring. These technical targets are related to a greater number of applications. Examples are information storage layers, films with quantum effects, optical layers, multilayer piles for quantum/semiconductor laser and x-ray optical compounds, displays, sensor layers, tribologic films, biocompatible films, photovoltaic films, membrane films, and chemically active surfaces (Bachmann 1998).

One might find that activities within the context of this *leitbild* follow a generalization pattern where promising targets are used to develop promising techniques. Realized techniques permit already sufficiently exact operations at the nano level to convey an idea of future products. The idea of future products, in turn, requires even more exact and precise tools.

4. Bottom-up techniques. While the preceding *leitbild* is still to be positioned in the field of top-down nanotechnology, the subsequent *leitbild* addresses the nanoscale from a different, "bottom-up" perspective. This perspective attempts to simulate nature to develop materials with novel properties by way of self-organization. These approaches are often dubbed "biomimetics" (topics 19, B).

In this case, an integration of various techniques is necessary. It has been shown that one can create structures by way of self-organization in a biomimetic process. However, our technological means are inadequate to fully utilize the potential this *leitbild* offers us. Being aware of the general feasibility — thanks to already realized artifacts — we can make reasonable assumptions about the requirements of the techniques necessary to pursue this path of development further.

5. Direct atomic control. Another *leitbild* focuses on direct control of atoms, to re-arrange them into new atomic structures (topic 15). This could lead to novel materials. The difference between the materials approach mentioned above and this one is related to how one controls the process (bulk reactions *vs.* atomic control). Atomic control is also strongly related to the idea of atoms being effectively used as carriers of certain functions, such as data storage, etc.

Table 6.8 gives an overview of other nanotechnology-related studies.

Table 6.8. Other Foresight and Technology Studies on Nanotechnology

Country	Activity
Germany	• Delphi study and report on "Technologies of the 21st Century" (Ministry for Education and Research) • study of VDI-TZ (Technology Centre of the German Engineers Society)
UK	• Parliamentary Office of Science and Technology (POST) report on nanotechnology (for the UK parliament)
The Netherlands	• Dutch Foresight Committee (OCV), studies on nanotechnology • Study Centre for Technology Trends (STT)
EU	• Scientific and Technical Options Assessment (STOA) study on nanotechnology (for the European Parliament) • IPTS studies on nanotechnology (European Commission)

Source: Adapted from ESANT 1999.

Societal Implications of Nanotechnology

The workshop's efforts are chiefly directed at identifying areas where mature nanotechnology can influence society. In Europe, the ESANT working group has undertaken the effort to outline possible avenues of societal implications of nanotechnology. The idea was to "foresee some beneficial aspects of nanotechnology for society in general" that reach beyond mere economic benefits, a topic a number of studies focus on currently. Table 6.9 gives an overview.

Table 6.9. Benefits

Aspects	Expected Benefits
Ecological	Small nano-sized particles have extremely high surface area compared to their volume; this property presents options for the fabrication of: • New catalysts • Heat reflection layers • Aerogels for transparent damping layers in solar architectures • Super thermal insulators • Transparent layers showing a higher resistance against wear and abrasion or anti-damping properties • Coatings to reduce wetting or dust adhesion so that windows can be cleaned by the rain • Special wall coverings for use in train carriages or on walls that can reduce the adhesion of "graffiti" so drastically that it can easily be wiped off
	• Magnetic nano-fluids as a cheap replacement for hydraulic oil in vibration damping systems and working machines, and as an abrasive and polishing medium for glassy materials • Selective colloidal membranes show possibilities for purifying waste water, removing dangerous waste and possibly bacteria as well, leaving only clean water of high quality
"Dematerialization"	• Nano-crystalline particles, with a mono disperse size distribution, to be formed into macroscopic parts with a higher strength and resistance against mechanical and thermal load, despite the smaller amounts of material required. These parts can be hard and flexible in a unit and can replace scarce materials. • New processing techniques, using remarkably lower temperatures, offer possibilities for minimizing energy consumption during component fabrication.
Clean technology	• Miniaturization in micro-electronics and precision technology and design of nano-catalysts, improving the efficiency and specificity of chemical reactions • Design of light and strong materials that lead to savings in energy and raw materials, especially in the transport sector • Applications in solar energy products

Table 6.9. Benefits (continued)

Aspects	Expected Benefits
	• Small, compressed particles enable new photovoltaic cells, with simpler structure than conventional ones • Plastics to be used as the electrode materials
Health	• More effective pharmaceuticals with reduced secondary effects due to improved basic understanding of the efficacy of natural human substances, like insulin or hormones • New form of localized drug delivery systems based on the potential of water soluble, pharmacologically active substances when attached to nanometer sized particles, also based on self-organizing hollow spheres that envelop the pharmacological substance • External control and incorporation of target information through incorporating magnetic particles or antibodies into the drug delivery system • Magnetic particles as option for cancer treatment by hyperthermic therapy • Preventive medicine could be made cheaper by using combinatorial systems of molecular surfaces in biosensors, while applications are possible in the military field, for example as premature warning systems for gas attacks. • Small machines in the blood stream or in the lymphatic circulation, to search for viruses, cancer or fat cells and to render these sickening agents harmless. Future applications for such machines are molecular repair processes in cells. • Nano-analytical properties also contribute to the detailed study of the synthesis steps during the development of pharmaceuticals.
Process security	• Quality assurance test systems: nanometer scale will become the precision standard for material analysis, control purposes and material treatment. – Magnetic storage disks – Electronic multilayer systems – Industrial polishing processes • New magneto-resistant multilayer systems offer drastically better positioning and controlling properties of sensors in automotive demands and as measuring systems for velocity, strain or work piece positioning.
Communication technology and electronics	• New production techniques that, for example, integrate the 3D architecture of storage systems, require more development attention • Vision of self-organized production of nano-sized electronic data storage and processing systems • Logical building blocks, of digital electronic units for example, based on particles or molecules • Molecular mechanical arrangements exhibit options for logical actions.

Source: Adapted from ESANT 1999.

NST is characterized as generic, having applications in more than one sector. An example is novel analytical techniques having similar applications in the environmental industry (e.g., for determining water quality), in healthcare (in the analysis of blood and blood products) and in the pharmaceutical and chemical industries (for testing for contaminants). Consequently, expected benefits are foreseen in a variety of areas. There are beneficial ecological aspects associated with nanotechnology. Another area is dematerialization. The notions of clean technology and production processes will become linked with further developed nanoscale technologies. Similarly, nanotechnological developments will be of great use in the health sector. Finally process security issues and communication technology and electronics are listed as areas of beneficial implications of nanotechnology.

Even though the ESANT study stresses the beneficial role nanotechnology is expected to play in economic and societal development, the group felt it is necessary to point to a possible detrimental effects and risks that may be associated with nanotechnology, a summary of which can be found in Table 6.10.

Table 6.10. Risks and Possible Detrimental Effects of Nanotechnology

Aspect	*Risks*
Artifacts based on nanotechnology incorporate genetic material or have genetic modification or repair as an objective	• Those risks that are associated with genetics, whether these are related to plants or animals (including human beings) • If the artifact incorporates some kind of computing and sensing element, say for the controlled delivery of a drug, additional risks arise for the patient if these elements should malfunction.
Physical scale of artifacts based on nanotechnology	• Invasion of privacy and of the human body through the planting and implanting of computing cum communication devices without those affected knowing this has been done • Ethical risks arise when implants of any kind are made, even more so if they are active with respect to the host, particularly human beings. Unlike macro-implants that can be inserted and removed by surgery, nano-sized implants will need very different approaches. • Security and the safety of the person, since it will be very difficult initially to detect the presence of nano-sized artifacts that are capable of breaching security and harming the individual.

Table 6.10. Risks and Possible Detrimental Effects of Nanotechnology (continued)

Aspect	Risks
Physical scale of artifacts based on nanotechnology (cont)	• An accentuation of environmental degradation through the uncontrolled spread of "waste" nano-sized artifacts (this has been referred to earlier as one of the possible consequences of disassembly associated with the notion of 100% recyclability). • In warfare, controlled distribution of biological and nerve agents may become feasible.
Ultimately self-assembling and self-replicating nature of nanotechnological processes	• Influence on employment opportunities in manufacturing need careful speculation. Processes of this kind are likely to be characterized by the need for a small number of very highly skilled people, exacerbating an existing trend in manufacturing industry. In addition, many processes will need to be redesigned embodying new principles, particularly relating to containment of active or waste products.
Nano-composites	• The general problem with composite materials is that they are more difficult to recycle and consume more energy during recycling than pure materials (e.g., coated glass versus untreated glass). • Given the present state-of-the-art of recycling technologies, wide-scale introduction of nano-composites is likely to increase environmental problems.

Source: Adapted from ESANT 1999.

To some extent, parallels are drawn with biotechnology, even though it is pointed out that that broad-brushed, general comparisons won't do justice to the technological field. Nanotechnology may lead to artifacts that incorporate genetic material or have genetic modification or repair as an objective and in this context, risks associated with genetic engineering may be of relevance also in terms of nanotechnology. Other areas that may involve some risk that should be taken into consideration are implications of the physical size of artifacts that are based on nanotechnology. A further group of risks is related to the self-assembling or self-replicating nature of foreseen nanotechnological processes. Finally, nano-composites were viewed as potentially increasing environmental problems.

The group points to the generic character of nanotechnology that *"places it in the same position as only two other technologies in recent decades, genetics and information technology. Of the latter two only genetics has received any extensive examination in terms of desirable applications and risks; information technology has escaped this kind of critical examination completely. Since nanotechnology embraces the entire set of genetics and information technology it is of paramount importance that its risks be speculated about in advance of desirable implementations, but not in an uninformed way"* (ESANT 1999).

The role of the media and a better understanding of this process was viewed as a critical factor. Furthermore, it was pointed out that *"methods of assessing risk, particularly those relating to toxicity, may themselves be inadequate to the new situations posed by applications of nanotechnology in many different fields."* If this was indeed to be the case, one should be aware of the time that could elapse *"before this [inadequacy] was recognized, allowing public perceptions of the risks involved to influence policy formulation in ways that would be difficult to reverse"* (ESANT 1999).

Therefore, it is important to question the judgements made and assess the process that has generated them in terms of its validity. This goes beyond specific techniques. Depending on the process and various other aspects, the "quality" of the judgements can vary substantially. One might call this "quality" of statements their *"epistemic value"* (Kuusi 1999b). Generalizations, or projections of certain technological developments, as well as judgements of their societal implications should be exposed to a critical discussion of their epistemic value. There is some, yet unpublished work going on in Europe that addresses these aspects and attempts to develop a methodology to assess the epistemic value. The work is carried out at a general level, but can also be applied to the context of NST. The gist of the concept is summarized in an equation. From the point of view of some actor k, one can describe the epistemic utility value ΔU_k as an evaluation of a technological generalization, or societal implications, as follows (Kuusi 1999b):

$$\Delta U_k = U_k(I^1 ,F^1 ,V^1_k ,R^1_k , L_k) - U_k(I^0 ,F^0 ,V^0_k ,R^0_k ,L_k))$$

The idea in the formula is that the epistemic utility of an evaluation for an actor depends on the value of five components after $(I^1, F^1, V^1_k, R^1_k, L_k)$ and before $(I^0, F^0, V^0_k, R^0_k, L_k)$ the presentation of the realized technology generalization/argument.

It is typical that the anticipated implications (I) of a topic considered in the bootlegging stage are not identical with the realized implications in the bandwagon stage. If a realized generalization/argument changes the assumed implications, I^1 and I^0 differ. An argument might also change the suggestion concerning the feasibility (F) of the used techniques, which means that F^0 and F^1 differ. The formula supposes that all relevant actors k understand the suggestions concerning implications and used techniques in the same way. Based on this assumption F and I are not related to specific actors k.

The differences of actor opinions are focused on the validity (V) and on the relevancy (R) of suggestions. It is reasonable to assume that in the bootlegging stage the evaluations concerning the validity differ considerable both inside the group of protagonists and especially between protagonists and outsiders. The proponents of the old paradigm do not believe that the proposed techniques could produce the proposed implications, or they believe that beside proposed implications the techniques produce other questionable implications. Opinions might also differ considerably concerning the relevancy of produced implications.

R^1_k and R^0_k are the relevancy of the topic or the technology generalization to the actor k after and before the presentation of the realized generalization/argument. If we consider the suggestions concerning future action, the reasonable realization of a topic depends on the other choices open to k. L_k describe these other choices. The model assumes that the realized technology generalization/argument has no implication on the epistemic value of competing choices based on other technologies

The research presented here is a theoretical model. It might be difficult to calculate such an "epistemic value" in practice. However, the underlying idea of the model is of great relevance. Contributors to studies of societal implications of NST must be aware of the varying validity and relevance of their assumptions and should address this issue in some structured way.

Part II: A Bibliometric Analysis of On-Going R&D Activities in Nanoscience and Technology

Part II of this paper will introduce a different way as to how one can approach nanoscience and technology from a social scientific viewpoint. This approach is widely quantitative, drawing on bibliometric and patent data. The study that is to be introduced here can illustrate the possibilities of social science research on current NST. It will also show the limitations of such approaches and identify areas for dialog between natural scientists and technologists on the one hand and social scientists on the other.

Social science might help scientists and technologists put their work into a broader context. The quantitative approach we are going to introduce here will not be able to tell us where NST will be in 25 years from now, but data like ours can be used to identify scientific and technological trends over a period of up to ten years. The data can convey an impression as to where nanotechnology originated and indicate the general direction in which the area moves. The bibliometric analysis we are going to present can also be used to classify the field itself and indicate to what extent work related to nanotechnology takes place in a disciplinary context. It can also help track the connection of science to patents.

However, one must be aware of the limitations of such approaches. They include in particular the key words approach used in data retrieval and the delineation of the subfields of NST. Publication and patent data was retrieved using a key word approach. These keywords always include some subjectivity. We used terms that were applied originally by bibliometricians, not experts in the field (see Table 6.17 in the Appendix for a list of key words). It should also be mentioned that the publication and patent subfields were delineated on the basis of established classification systems. These systems are structured after disciplines and technological sectors. This might not always be the most appropriate form of organizing a field-specific database. Finally, there are general methodological limitations associated with the tool of patent citation analysis, a summary of which can be found in Table 7A.4 of the Appendix.

In order to analyze knowledge generation in nanotechnology, we set up three different databases:

- A bibliometric database of all scientific research papers related to nanotechnology

- A database of all U.S. patents in that technological area

- A database that connects both data sets by identifying the references citing from nano-patents to nano-papers

While the bibliometric database allows us to have a look at how transdisciplinary current activities in nanotechnology-related sciences are, the second (patent) database offers a perspective at what has been done in terms of technological development so far. The third database will be instrumental in determining the extent to which scientific knowledge generation is related to its specific "application context," namely nanotechnology. We use the number of patent citations as an approximate measure for relevant connections.

1. Bibliometric Database. The bibliometric database contains 5,000+ scientific research papers on nano-related subjects. Following Braun et al., we retrieved all publication titles of journal articles in the Science Citation Index and patents downloaded from the USPTO Internet database that included the term "nano" as such or as a prefix. Subsequently, irrelevant records that contained only terms like nanogram, nanosecond were deleted. The Appendix contains a list of search terms.[8]

The publication data is analyzed at various levels, one of which is distribution by major and sub-disciplines.

Fig. 6.28 illustrates the paper distribution according to major and sub-disciplines. We use the SPRU classification scheme. This scheme aggregates journals of the science citation index according to their broad disciplinary orientation. One major finding is that the majority of the papers (71%) are published in a disciplinary context, even though the share of cross-boundary publications is relatively high.

Looking at the percentage distribution of nano-papers one can establish some trends. We calculated a linear slope coefficient based on the annual shares of major fields as well as subfields (see Appendix - Growth columns in Tables 6.18 and 6.19). While natural and multidisciplinary sciences have gotten stronger, the major fields of engineering and materials and life sciences seem to have lost importance. Papers in natural science have a slope coefficient of 3.2. The only other growing field is multidisciplinary sciences with 0.4. Engineering and materials as well as life sciences lost ground dramatically. Their shares dropped by more than a half and three fourths, respectively, leading to coefficients of -2.3 and -1.5.

A look at the subfield results gives a more detailed view. There we see that the growth of nano-papers with a natural science classification is not uniform. Thus the slope of chemical nano-publications is twice as steep as the one of physical nano-papers.

---

[8] More detailed information regarding the data retrieval can be found in Meyer and Persson 1998 and in Meyer 2000.

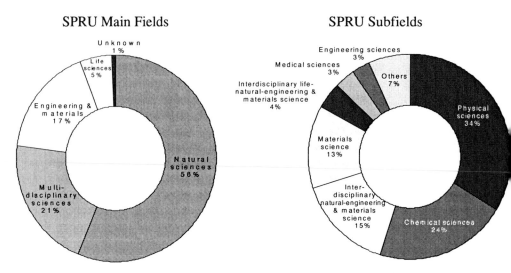

Figure 6.28. Nano-publication database: disciplinary distribution.

The growth in share of nano-papers in the major field of multidisciplinary publications is due to the subfield of interdisciplinary natural-engineering and materials science. While the shares of other multidisciplinary sciences are decreasing, this subfield could increase its share by more than a third. Its slope coefficient equals 1.3. An interesting observation in this connection is the decrease in share of publications in materials science (-1.9). This more than what happened to the other multidisciplinary fields seems to have contributed to the growth of that particular subfield.

2. Patent Database. The patent database comprises approximately 2,600 patents, mainly in instrumentation, electronics and electrical engineering, and chemicals/ pharmaceuticals. Table 6.19 shows the development of the database over time and the distribution of patenting after technological fields. One can recognize a clear focus on instruments, electronics, and chemicals/pharmaceuticals. An interesting trend one can make out is the strengthening of the chemicals/pharmaceutical sector. An update of the database (done in 1999) would show if this trend prevails. As the USPTO will make available patent application data in the near future, the analyses should be significantly improved since the time lag will

Table 6.11. Distribution of Nano-Patents Over Time and By Technological Fields

Period filed: Sector	1969	1972	1973	1974	1975	1976	1977	1978	1979	1980	1981	1982	1983	1984	1985	1986	1987	1988	1989	1990	1991	1992	1993	1994	1995	1996	1997	1998	Grand Total
Instruments	1		1	5	6	8	12	10	11	15	14	21	23	19	12	26	27	24	35	38	48	59	61	77	107	76	27		763
Electronics and Electr. Mach.			1	5	5	7	9	11	10	10	13	13	11	14	15	21	17	18	28	30	41	55	68	93	109	69	22		696
Chemicals/Pharmaceuticals			1	1	11	13	5	4	4	4	7	6	4	5	5	9	11	10	15	17	25	41	48	77	143	65	29	1	561
Other machinery			2		2			4	3	5	1	2	3	2	5	6	4	8	9	10	13	17	23	21	40	32	5	1	219
Metallurgy				1		2	1	2	1	4	1		3	3	2		2	6	7	2	11	15	10	17	36	19	10		154
Other industrial products										1	1		1	1	1	3	3	2	3	3	12	6	9	21	18	23	2		110
Food										1		1	2			1	1			1	3	1	3	6	6	8			34
Other						1		1	1	1				2		1		1	2	2		2		2	3	1			19
Textiles		1					1	1							2					1			3	3	8				19
Nucleonics		1			1	1			1				1		1		1		1	3	1	2	1	1	1	1			14
No classification					1	1													1	2	2	1		1	1	1	2		11
Transport								1						1	1					1			1		3				10
Paper, printing and publishing												1			2				1	1		2	1				1		8
Motor vehicles																			1	1			1						3
Oil refining																				1					1		1		3
Aerospace																						1							1
Construction																										1			1
Grand Total	1	2	5	12	26	32	28	33	30	41	37	44	48	47	44	69	66	69	102	112	157	201	229	320	475	296	98	2	2626

be minimized. It might be useful in terms of the societal implication efforts to have a closer look at a number of areas, trying to identify paths of technological development and subsequent industrial applications. Based on these trends, one could develop scenarios of potential societal implications these applications might have.

As these sectors are generally attributed a certain proximity to the science-base, one should expect a substantial overlap between nanoscience and nanotechnology, indicated by a considerable number of relevant nano-research papers being cited by nano-patents.

3. Patent Citation Database. The patent citation analysis confronted us with somewhat surprising, counter-intuitive results. One should assume that a technological field that is generally acknowledged as science-based[9] would encompass patents that frequently cite the corresponding set of scientific research papers. This, however, is not the case. Matching the 5,000 plus nano-papers identified in the Science Citation Index (1991-96) with the nano-patents resulted in 275 matches.[10] A test matching procedure, with 22,000 nano-related papers found in the INSPEC database, resulted in 371 matches. Given the size of the source databases, the relatively small number of corresponding citations can be taken as an indicator of weak interaction between nanoscience and technology.

Figure 6.29 gives an overview of the results of the matching procedures.

Linking scientific and technological databases can go beyond evaluating the intensity of exchange processes in an emerging field. It can also help illustrating and understanding if and how new developments in science and technology might find their way into society. We can follow up how technological sectors are linked to certain domains of science in NST. This again can be specified for various industries and organizational channels.

---

[9] One illustration of this is, for instance, a major EU-funded conference on 'Nanoscience For Nanotechnology'; the title implicitly assumes that nanotechnology builds critically upon nanoscience.

[10] It should be noted that the matches we found are patent citations listed in the 'other references' section on the front-page of the patent, referring to scientific papers in our nano-publication database only. This methodology corresponds to current practice amongst patent bibliometricians who track front-page citations of non-patent literature to study the science/technology linkage. The following section on the nature of patent citations will clarify the importance of this practice.

This again may allow actors to formulate educated guesses about future developments.

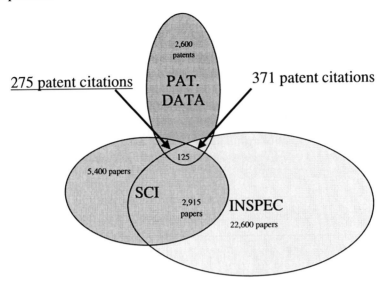

Figure 6.29. The different data sets and their overlap.

If one looks at nano-patent citations by industrial sectors and SPRU-main classes, the instruments sector appears to be the sector that relies on nanoscience the most (with 49 citations). Instruments are followed by electrical machinery (40 references), electronics (35), pharmacy (29), and chemistry (21). Other sectors refer to scientific domains less than 20 times. The most referenced scientific domain is natural sciences (145 citations), followed by multidisciplinary sciences (87).[11]

---

[11] Given the dominance of natural sciences it might be interesting to have a closer look at the level of subclasses. The subfields that were cited ten times or more are: physical sciences (with 108 citations), multidisciplinary sciences (40), interdisciplinary natural-engineering and materials (36), chemical sciences (34), materials science (19), medical sciences (14), and inter-disciplinary life-natural (10). The data indicate clear differences in the linkage patterns of the industries investigated. While instruments, electric machinery, and electronics have a strong focus on physical sciences (with 19 out of 49 nano-patent citations, 23/40, and 24/35, respectively), the pharmaceuticals and chemicals industries have different orientations. For instance, with ten out of its 29 citations, the pharmaceuticals industry is linked to the medical sciences much stronger than any other industry. Similarly, it is connected to inter-disciplinary life-natural by 8 patent citations. In terms of nanoscience and technology, the chemicals industry seems to put emphasis on inter-disciplinary natural-engineering and materials and multidisciplinary sciences to a relatively greater extent than instruments, electric machinery and electronics industries. Various combinations of these and other sets of indicators confirm the results presented here. For instance, the distribution of patent citations as measured by Dewey descriptors and IPC classification at the 4-digit level present similar results. Physics/engineering with 66 citations is the major 'contributing' scientific domain. The strongest connection has been established with subclass H01L

Table 6.12 gives an overview of the distribution of the patent citations. Not only is the overall interaction between nanoscience and nanotechnology relatively weak with 275 citations, but almost two fifths of these citation links are between papers written at public sector and research organizations (PSR, including universities, research establishments, etc) and patents held by PSR exclusively. Less than 30 % of the 275 patent citations actually connect work of a PSR author and an assignee in industry.

Table 6.12.   Patent Citations By Organizational Categories

Patent Assignation / Author Affiliation	University	Multinational Corporation	Small and Medium-Sized Enterprise	Industrial Association	Other	Total
University	68	34	23	5	8	138
Multinational Corporation	8	21	8	4	6	47
Research Establishment	13	5	9	5	5	37
Multinational Corp. and University	4	6	3	1	3	17
University and Research Establishment	7	2	3		4	16
Small and Medium-Sized Enterprise	1	2	2	2	0	7
Other	1	3	3	1	4	12
Unknown	0	1	0	0	0	1
Total	102	74	51	18	30	275

(20 citations). Other links are with H01J (6 citations) and G03F (4 citations). The second biggest contributing 'science' is the field of comprehensive works and general sciences with 37 references, which are much more evenly distributed. With 6 citations, H01J tops the list of technological domains linked to this field of science. Runners-up are H01L and C01B. The scientific field 'physics' holds rank #3 with 25 patent citations. However, there is no technological field that more than 3 citations are related to. The Dewey-domain 'engineering/chemical engineering' is the fourth biggest attractor of nano-patent citations, together with 'pharmacology and pharmacy' (both 15 references). These two areas, however, relate to nano-technologies in entirely different manners. While (chemical) engineering is referenced just by B32B four times and linked to eight other technological domains with one or two patent citations, 'pharmacy and pharmacology' are clearly linked to subclass A61K (with 12 out of 15 patent citations). There are three further linkages established by individual patent citations (A61F, C07F, B01J; one citation each). The only other 'stronger' link is the connection between the scientific field of instruments and the IPC subclass H01J with 4 references.

The following tables present a number of examples arranged by field and organizational category. Thereby we can gain a perspective on how patent citations describe the science/technology interface in organizational terms for a specific subfield.

The case of nano-patent citations in the field of *"other machinery"* (Table 6.13) illustrates a reference pattern in which nanoscience-citing university nano-patents substantially rely on scientific nano-publications authored in the university system or non-industrial research centers. Similarly, it also shows that respective patents assigned to multinationals chiefly refer to work from their own organizational category. The case of *metal products* (Table 6.14) points to the universities as the main producer of nano-scientific information relevant to nano-patents. However, it also demonstrates the universities are the most citing assignee organization in this area. This might raise questions as to how "industrial" academic patents are. The case of *electronics* (Table 6.15) illustrates the importance of universities as science producers. However, university-held patents cite university-generated research much more frequently than others. Industry, in the form of multinational corporations, appears to be an important producer of scientific knowledge and also a major absorber. The case of *nano-instruments* (Table 6.16) indicates the importance of intermediary organizations in terms of knowledge diffusion. Industrial associations and similar organizations seem to be the second largest absorbers of knowledge in the field. Again, it can be pointed out that university-papers are most frequently cited in university patents.

Table 6.13. Other Machinery

Patent Assignation Author Affiliation	Multinational Corporation	University	Industrial Association	Grand Total
University	1	2	2	5
Multinational Corporation	3			3
Research Establishment	1	2		3
Multinational Corporation And University	1			1
University And Research Establishment		1		1
Grand Total	6	5	2	13

Table 6.14. Metal Products, e.g., Machines

Patent Assignation / Author Affiliation	University	Multinational Corporation	Small and Medium-Sized Enterprise	Grand Total
University	8	1	1	**10**
Multinational Corporation	3	1		**4**
Research Establishment	3	1		**4**
Small and Medium-Sized Enterprise	1			**1**
Grand Total	**15**	**3**	**1**	**19**

Table 6.15. Electronics

Patent Assignation / Author Affiliation	University	Multi-National Corporation	Small and Medium-Sized Enterprise	Government Agency	Research Establishment	Grand Total
University	11	4	2	1		**18**
Multinational Corporation		4	3			**7**
University And Research Establishment	2	1		1	2	**6**
Research Establishment		1		1		**2**
Multinational Corporation And Government Agency				1		**1**
Small And Medium-Sized Enterprise		1				**1**
Grand Total	**13**	**11**	**5**	**4**	**2**	**35**

302

Table 6.16. Instruments

Patent Assignation / Author Affiliation	University	Industrial Association	Multi-National Corporation	SMEs	Research Establishment	Government Agency	Grand Total
University	13	1	6	2			22
Multinational Corporation	3	3	1			1	8
Research Establishment	3	3					6
University and Research Establishment	5						5
Multinational Corporation and University	1	1	1	1			4
Small and Medium-Sized Enterprise		1		1			2
Hospital		1					1
Multinational Corporation and Research Establishment					1		1
Grand Total	25	10	8	4	1	1	49

These examples illustrate insights that can be gained from the application of patent citation analysis at the combined sectoral/organizational levels. We could demonstrate that the NST cluster is very heterogeneous and its different subfields show rather varied patterns of knowledge transformation, diffusion and absorption.

Outlook

What can we learn from above analysis? First, we have realized that science related to the nanoscale takes place mainly in a disciplinary context, and, therefore, we are far away from a unified nanoscience. However, we were able to point to a relative increase of natural scientific and multidisciplinary publications and a relative decrease of materials sciences. The patent database we set up illustrated the broad range of technological artifacts and processes where the nanoscale matters. An analysis of the citation linkage between nano-patents and nanoscience indicated that there is just a weak direct connection between technology and science at the nanoscale. Moreover, an investigation of patent citation has shown that many of the patents that do cite nano-papers are assigned to universities. This could indicate a process of knowledge transformation from science to technology within the academic sector rather than a knowledge transfer process from academe to industry. Further analysis also pointed to sectoral differences as far as who cited whom. There were different organizational patterns to be observed.

The data should have illustrated that NST will develop various aspects in many, different ways. And the citation analyses also should have shown that each of the areas NST can affect follows different transformation and diffusion patterns. Studies on the implications of NST for society will need to pay attention to this complexity.

The tool introduced is a good method to map activities in NST. At more advanced stages it can also be used to track more specific elements. Using such a social scientific methodology on a NST can also be instrumental in facilitating a basis for a common language. The results of this study could create a discussion about delineation and character of NST and their various sub-fields. This discussion will familiarize natural scientists and engineers with social scientific thinking while it forces social scientists to deal with scientific and technological aspects of the heterogeneous and complex field of nanoscience and technology.

This mutual understanding is of fundamental importance if one is to study societal implications of nanotechnology. Social scientists need to gain an understanding of where the new technology is coming from to be able to discuss potential societal implications of nanotechnology in a meaningful manner. This approach also would counteract a simplified understanding of mature nanotechnology as the realization of the assembler visions of the molecular manufacturing pioneers. The major point this paper wants to make is that societal implication studies of (mature) nanotechnology should not start with unsubstantiated visions of the future but begin with an effort to understand NST as it emerges.

The social implications program of workshops provides the chance to start a dialog between scientific and industrial experts and social scientists in the area. Social scientific research like the contributions presented could form a platform to build on. Some of the European studies introduced in Part I underline the point made at the workshop that in particular areas it makes sense to analyze past experiences with other technologies to prepare for future experiences with nanotechnology. Biotechnology is an apparent candidate for this undertaking. However, one must be aware of the risks of unspecific comparisons.

Finally, another area that needs to be addressed is the process of evaluating and assessing potential societal implications itself. European studies on NST pointed to the need for measures that can control reliability and validity problems that come along with implication research on future technologies. It is important to find a solution for this issue or at least a way of communicating its implications if one wants to ensure that the implication information generated will be used in an appropriate manner.

Appendix

Table 6.17. Search Terms Used by Braun et al.

All terms searched begin with nano:

~crystalline, ~structure, ~particle, ~scale, ~composite, ~tube, ~crystal, ~gram, ~meter-Size, ~phase, ~cluster, ~size, ~capsule, ~crystallite, ~sphere, ~flagellate, ~metric, ~filtration, ~lithography, ~fabrication, ~indentation, ~technology, ~colloid, ~porous, ~wire, ~bridge, ~crystallization, ~tubule, ~electronics, ~vid, ~particulate, ~tribology, ~foam, ~diffraction, ~tip, ~aggregate, ~crystallized, ~flare, ~material, ~dispersed, ~filament, ~powder, ~rheology, ~architecture, ~layer, ~lithographic, ~channel, ~device, ~electronic, ~fiber, ~granular, ~heterogeneous, ~meter-Thick, ~peptide, ~space, ~-Y-TZP/Al2O3, ~droplet, ~feature, ~gold, ~grained, ~mechanics, ~multilayer, ~pore, ~probe, ~whisker, ~analysis, ~ball, ~cavity, ~characterization, ~column, ~constriction, ~disperse, ~dispersion, ~electrode, ~fibrillar, ~manipulation, ~precipitate, ~radian, ~system, ~technique, ~template, ~topography, ~world, ~analytical, ~band, ~cell, ~cermet, ~chemical, ~chemistry, ~coll, ~crack, ~cyrstalline, ~diamond, ~dissection, ~domain, ~engineering, ~friction, ~grain, ~granule, ~hardness, ~heterostructure, ~laminate, ~layered, ~machined, ~mechanical, ~metal, ~meter-Particle, ~meter-Thickness, ~molecular, ~photonics, ~physics, ~porosity, ~processing, ~replica, ~rod, ~science, ~titanate, ~twin, ~vision, ~wear, ~-AMP, ~-Mg2Si, ~-NMR, ~-Na-15, ~composite, ~-SnO2, ~-Tin, ~-ZrO2, ~-object, ~-optical, ~amorphous, ~anatomy, ~anodization, ~apatite, ~battery, ~body, ~building, ~cage, ~capillarity, ~capsular, ~carbon, ~carrier, ~catalysis, ~ceramics, ~climate, ~compound, ~conductor, ~cone, ~construction, ~contact, ~crystal-Doped, ~cube, ~cyclitic, ~deformation, ~dimension, ~dimensional,~disk, ~dislocation, ~displacement, ~drop, ~droplet-ABA, ~dynamical, ~elastohydrodynamics, ~electromechanical, ~electron, ~element, ~encapsulated, ~environment, ~equivalent, ~etching, ~fabricated, ~fibril, ~filtered, ~gate, ~gauge, ~glass, ~heterogeneity, ~heterotrophic, ~hole, ~inclusion, ~ionics, ~junction, ~kinematics, ~laminated, ~lithographically, ~machine, ~machining, ~magnetism, ~mask, ~matrix, ~mechanism, ~mental, ~meric, ~meter-Structure, ~meter-T, ~meter-Thin, ~meter-Width, ~metersized, ~modification, ~optics, ~pattern, ~pinhole, ~pipe, ~pit, ~polar, ~polyhedra, ~porous, ~positioner, ~precipitation, ~programmed, ~reaction, ~reactor, ~rheological, ~roughness, ~scaffolding, ~scatterer, ~sled, ~slider, ~solid, ~source, ~spacing, ~spectroscopy, ~strain, ~strip, ~subharmonics.

There are several ways that nano-science and nano-technology are related to each other and their environment. Nano-science can provide relevant documents that are cited in patented nano-technology (arrow 1). In a similar way, other scientific papers might be cited in nano-patents (arrow 3). In addition, nano-technology can contribute to nano-science. In some instances, this might lead to citations of patents in the respective literature (arrow 2). Also patented technological developments beyond the nano-scale might be of relevance to nano-scientific research, which might subsequently refer to it (arrow 1). It should be pointed out that in this article we 'weave' a cognitive web between nano-science and nano-technology only. Therefore we just track linkages that are illustrated by arrow 1. We neglect all other interconnections (as depicted by arrows 2-4).

This is due to the nature of patent citation analysis and the restricted availability and access of bibliometric data. For instance, it was not possible to get access to a national publication database in a large country. This way it would have been possible to track all scientific contributions to the sample of nano-patents.

The citation analysis was carried out by matching the two databases. For the nano-publications, search keys based on author names, journals and publication dates were defined. Those were then searched for in the patent database. It should be noted that this study analyzes front-page patent citations only. This is due to the fact that patent citations in the specification part of the patent document are usually not made available in a bibliographic database.

Patent citations and their context.

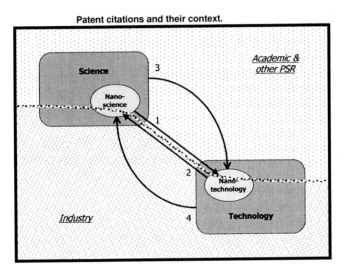

The matching procedure enables us to determine the overall overlap between nano-patents and papers and allows us to set up a database of patent citations. For each patent citation, the databank provides information about

- title, inventor name, inventor address, assignation, and technical classification of the citing patent, and
- title, author name, author affiliation, and journal classification of the cited article.

The database will provide the basis for our citation analyses. Firstly, we analyze patent citations by technological sectors and scientific domains. Then, we investigate patent citations by author affiliation and patent assignation. In a final step, we combine both types of analysis by looking at the organizational distribution of patent citations in a number of technological subfields.

Figure 6.30. Limitations of the patent citation approach & methodological remarks.

Table 6.18. Distribution of Papers by Major Field by SPRU Classification

Major Field	1991	1992	1993	1994	1995	1996	Total	Growth[a]	Standard Error
Natural Sciences	119	205	386	612	765	946	3033	3.2	3.7
Multidisciplinary Sciences	51	103	141	189	289	367	1140	0.4	2.2
Engineering and Materials	74	92	127	184	235	209	921	-2.3	1.9
Life Sciences	30	47	30	59	70	56	292	-1.5	1.8
Unknown[b]	0	0	0	1	5	24	30	0.2	0.4
Total	274	447	684	1045	1364	1602	5416		

a   The growth rate is calculated as a linear slope coefficient based on the annual shares of the respective major fields.

b   The SPRU-classification scheme is based on ISI 1994 journal set. ISI add journals and drop journals. Ten of the 30 unknown papers are in journals without ISI classification anymore; the remainder of 20 unclassified papers has been published in newly added journals.

Table 6.19. The Subfield Distribution of Nano-Papers by SPRU-Classification

Subfield	1991	1992	1993	1994	1995	1996	Total	Growth[a]	Standard Error
Physical sciences	74	132	266	387	420	548	1827	1.1	4.6
Chemical sciences	40	70	107	215	334	375	1141	2.2	1.8

Table 6.19. The Subfield Distribution of Nano-Papers by SPRU-Classification (continued)

Subfield	1991	1992	1993	1994	1995	1996	Total	Growth[a]	Standard Error
Inter-disciplinary natural-engineering & materials science	26	67	92	144	211	286	826	1.2	1.6
Materials science	57	75	99	133	185	153	702	-1.9	1.4
Inter-disciplinary life-natural-engineering & materials science	15	22	38	30	58	59	222	-0.4	0.9
Medical sciences	20	30	21	39	42	34	186	-1.0	1.0
Engineering sciences	14	12	18	33	39	39	155	-0.4	0.8
Inter-disciplinary life-natural sciences	10	14	11	15	20	22	92	-0.5	0.5
Biological sciences	7	7	7	12	15	10	58	-0.3	0.4
Interfield engineering and material sciences	3	4	8	18	10	14	57	0.0	0.4
Interfield natural sciences	2	3	10	8	11	21	55	0.1	0.3

Table 6.19. The Subfield Distribution of Nano-Papers by SPRU-Classification (continued)

Subfield	1991	1992	1993	1994	1995	1996	Total	Growth[a]	Standard Error
Interfield life sciences	2	9	2	8	11	10	42	-0.1	0.6
Earth sciences	2	0	3	2	0	2	9	-0.1	0.3
Information and communications	0	1	2	0	1	3	7	0.0	0.1
Agricultural sciences	1	1	0	0	2	2	6	0.0	0.1
Mathematical science	1	0	0	0	0	0	1	-0.1	0.1
Unknown[b]	0	0	0	1	5	24	30	0.2	0.4
Total	274	450	686	1047	1366	1607	5416		

a The growth rate is calculated as a linear slope coefficient based on the annual shares of the respective major fields.

b The SPRU-classification scheme is based on ISI 1994 journal set. ISI add journals and drop journals. Ten of the 30 unknown papers are in journals without ISI-classification anymore; the remainder of 20 unclassified papers have been published in newly added journals.

References

Bachmann, G. 1998. "Innovationsschub aus dem Nanokosmos." *Technologieanalyse*. Ed. By VDI-Technologiezentrum, Düsseldorf, Oct., section 2.3.2.

BMBF. 1996. "Delphi-Bericht 1995 zur Entwicklung von Wissenschaft und Technik – Mini-Delphi." Bonn, author's translation.

Esant. 1999. "Economic and Social Aspects of Nanotechnology." First draft report of working group 6 of the Euroconferences on Nanoscience for Nanotechnology. Also at http://www.nano.org.uk/ ESANT99.htm.

Frenken, J.W.M. 1998. "Scanning Tunneling Microscopy", section 5.2, A. ten Wolde (ed.), *Nanotechnology: Towards a molecular construction kit.* STT Report #60, The Hague, 289-299.

Kuusi, O., 1999. "Epistemic value analysis – a tool to evaluate technological options." Personal communication.

Kuusi, O. 1999b. *Expertise in the future use of generic technologies.* VATT Research Reports #59, Helsinki.

Kuusi, O. and M. Meyer. 2001. "Technology Generalizations and Leitbilds – The Anticipation of Technological Opportunities." Manuscript submitted to *Technological Forecasting and Social Change.*

Meyer, M. and O. Persson. 1998. "Nanotechnology - interdisciplinarity, patterns of collaboration and differences in application," *Scientometrics,* XXXXII (2), 195-205.

Meyer, M. 2000. "Patent citations in a novel field of technology: What can they tell about interactions of emerging communities of science and technology?" *Scientometrics,* XXXXIIX (2), 151-178.

Meyer, M. 2000b. "Hurdles on the way to growth. Commercializing novel technologies." Institute of Strategy and International Business, Working Papers Series 2000/1, Helsinki University of Technology, Espoo.

Philipse, A.P. 1998. "Colloidal Dispersions," section 3.4, A. ten Wolde (ed.), *Nanotechnology: Towards a molecular construction kit.* STT Report #60, The Hague, 171-178.

ten Wolde. (ed.). 1998. *Nanotechnology - Towards a molecular construction kit.* The Hague: Netherlands Study Centre for Technology Trends (STT).

NANOTECHNOLOGY AND UNINTENDED CONSEQUENCES

Edward Tenner, Princeton University

Radically new technology inspires lyrical utopianism and melancholy catastrophism. What is new at the turn of the twenty-first century is the note of alarm among the leaders of change. Eric Drexler and Bill Joy have focused on the self-replicating potential of nanotechnology processes. From the outset, Drexler has acknowledged the possibility of rampant synthetic organisms that could displace real ones (the "gray goo problem,") and Joy has even speculated that quasi-human robotic systems constructed with nanotechnology could in effect enslave our species. But there is old-fashioned optimism, too. Enthusiasts foresee agricultural bounty, a paradise of health and longevity, mental and physical enhancement, and a wonderland of novel consumer goods. The web site of a forthcoming lay publication, *NanoTechnology Magazine*, promises "friendly energy" and "positive impact agriculture," the end of animal experimentation, and the neutralization of all chemical and even radioactive waste (http://nanozine.com).

The history of technology can not reconcile these visions. But it can help prepare us for the surprises that have always been the result of human ingenuity. We can expect five things: (1) The experts will be seriously wrong about at least some important things. (2) Long-term, cumulative problems will be a greater problem than the perils of catastrophe. (3) Organizing and supervising nanotechnology will create dilemmas. 4) Successes may be as costly as failures. (5) We probably have not imagined the greatest benefits of nanotechnology, either because they seem too technologically modest or because they may result from improbable chains of events.

Expertise in the Long Term

The most gifted scientists and engineers have a mixed record as long-term forecasters. Lord Kelvin, outstanding as an inventor as well as physicist, is now also known for his prediction that heavier-than-air flight would remain impossible. Irving Fisher, equally eminent a generation later as an economist and entrepreneur, declared in 1929 that the stock market appeared to have reached a permanently high plateau. In a 1955 book, *The Fabulous Future*, published by *Fortune* magazine, John von Neumann predicted that by 1980 improvements in reactor technology would make nuclear energy so economical that it "may be free – just like the unmetered air — with coal and oil used mainly as raw materials for organic chemical synthesis...." Following earlier scientists, he did recognize the human influence on what later became known as global warming, but he also foresaw advances in atmospheric science leading to deliberate climate modification "on a scale difficult to imagine at present," including the possibility "of a new ice age or of a new tropical . . . age in order to please everybody." Yet von Neumann was equally far too modest in his expectations for the electronic computer. He imagined its future as ever more powerful central control of economic planning and industrial processing. His Cold War vision for the device he had helped to create had no place for the miniaturization and decentralization of electronics (von Neumann 1956).

Of course scientists and engineers imagine the forms their ideas will take and the consequences they will have. The operations of an agency like the National Science Foundation depend on projections of benefit to humanity

312

as well as abstract knowledge. But just as many writers are surprised by the uses to which their ideas are put by readers, innovators must be prepared to find that reality has other plans for their work.

From Acute to Chronic Risks

Unlike von Neumann, we know the outcome of the Cold War and how it surprised liberal and conservative forecasters alike. Nuclear energy is neither as threatening nor as immediately promising as expected. But our experience in the 45 years since von Neumann's essay suggests some lessons. In our imagination of technological consequences, we think first of disasters. But for the past century, our concerns have been misplaced. Technology has instead tended to replace catastrophic problems with chronic ones. In place of the threat of an apocalyptic thermonuclear exchange that would be over in days we face the dilemma of dealing with lethal nuclear wastes, some of which will last for hundreds of thousands of years. Freon coolant protected people from once-common refrigerator explosions, but at the price of slowly depleting the earth's ozone layer. Low back pain and cumulative trauma disorders like carpal tunnel syndrome now are more serious industrial hazards than loss of life and limb. Where effective treatment is available, AIDS has become not the new plague that so many feared in the 1980s but a long-term, controllable condition. Post-World War II reinforced construction techniques have generally been safe but not necessarily stable. Architects and engineers failed to predict the many interactions of materials, moisture, and weather. Rehabilitating these once advanced buildings has become a complex and costly specialty. It takes time but it, too, is manageable. We defuse problems by diffusing them.

There is no guarantee that the future will be like the past. But if the trend of recent decades continues, the hazards of nanotechnology will not be apocalyptic "gray goo, "uncontrollable self-replicating substances or organisms. Chemists and physicists in nanotechnology research argue persuasively that self-assembling "nanobots," drawing resources from the environment as living creatures do, will never be possible. Nanotechnology, alone or with other innovations, nevertheless might indirectly promote some existing pathogen or parasite. Early in the twentieth century many foresaw the uses of aircraft for war; far fewer realized how effectively mass civilian air transportation and vastly expanded, containerized seaborne commerce would inadvertently help new disease agents cross ocean barriers, including HIV and the West Nile virus. Likewise the spread of mad cow disease (BSE) as human Creutzfeldt-Jakob Disease (CJD) in England arose from the interaction of innovations: changes both in animal feeds and in methods of

313

extracting meat particles from carcasses. The Legionnaire's Disease bacillus has always been common in natural waters without causing epidemics; modern heating, ventilating, and air conditioning systems gave it rare, ideal conditions for reproduction and transmission. Risk analysis might show many technologies to be safe in isolation, yet the range of possible interactions remains beyond it.

Present testing techniques are not consistently able to identify slow, gradual change. Yet environmental damage is most likely to occur in some apparently innocent indirect result of an apparently benign process, just as we are now discovering the problems created by decades of the distinctive chemical wastes of electronics manufacture. One of the greatest challenges of any new technology is determining which new potential problems need to be identified and measured. Diesel engines, for example, produce relatively few pollutants associated with conventional internal combustion engines, but have unique emissions problems of their own. It took time to develop new, appropriate tests for them. R. Flagan and D.S. Ginley, in their chapter on "Nanoscale Processes in the Environment" in the NSTC report (NSTC 1999) observe that we still barely understand how nanoparticles and nanostructured materials affect living organisms and other aspects of the environment. Nanoparticles could set off subtle changes in plant or animal tissues that could cascade into extensive biological change, just as DDT did in the postwar years after it was hailed as an environmental breakthrough. And a series by Tom Horton and Heather Dewar in the *Baltimore Sun* during the week of 24 September 2000 reveals how the most important chemical innovation of the early twentieth century, the chemical extraction of nitrogen from the atmosphere, helped end famine and increased the world's population by an estimated two billion people — but has also slowly and indirectly choked the life of rivers and coastal waters by depleting oxygen. The leading authority on global nitrogen, the Canadian environmental scientist Vaclav Smil, has called the Haber-Bosch process that makes modern fertilizers possible the most important invention of the twentieth century.

Slow, unforeseeable processes may also affect engineering applications of nanotechnology. While smart concrete may be designed to counteract the corrosion that has beset the conventional product, how can it be formulated to anticipate environmental changes brought about during its lifetime by other new substances, including other new products of nanotechnology itself? The very diversity and excitement of nanotechnology research may paradoxically be a hidden weakness. How can we anticipate real-world interactions when we are going to be modifying reality on so many fronts, and when data can take time to interpret? A hundred years after the first

314

observations of global warming, there is still debate over how much of it is due to human causes. And how can we expect to reverse or even curb chronic problems when they take so long to become serious that we have become dependent on the products that have caused them? Nanotechnology may well help us, directly or indirectly, address nitrogen and other environmental issues, but it is equally likely to start changing things before we learn which changes to monitor.

Nanotechnology and Security

The social issues of nanotechnology will be at least as complex as its consequences for the environment and design. In his book *Normal Accidents*, the sociologist Charles Perrow has argued that some technologies are both nonlinear and tightly coupled. That is, they can feed on themselves *and* a single mistake can be readily transmitted through the system (Perrow 1984). Nanotechnology may reduce these forms of risk significantly, especially by promoting energy efficiency and making it possible to reduce or eliminate nuclear power production. But it is also possible that some otherwise extremely desirable nanotechnology processes may meet Perrow's definition. They may require the kind of strict controls that we now associate with the nuclear industry — controls not only on operations and physical access but also on the sharing of knowledge. To von Neumann's generation these appeared the inevitable costs of a national security state. In the early twenty-first century, as the problems of America's national laboratories show, this system of command is harder to sustain. Knowledge has become far less viscous. Laptop computers and Web connections are sabotaging decades of carefully defined security procedures, yet we are warned that stringent new security could wreck the morale of the producers of classified knowledge. Innovations flow readily across national borders because the sheer volume of world trade and communication overwhelms controls. The programmers in recent headlines have been Finnish, Swedish, and Belgian; tomorrow's may be from any part of the world. Proliferation of bureaucratic controls, extending in the United States to state and local levels, creates a new set of problems in the selection and supervision of controllers.

Although some forms of research will be clearly unethical, and already banned under treaties against chemical and biological warfare (CBW), the line often is not clear. Haber and Bosch originally developed their method for military purposes, yet until comparatively recently it was acclaimed as a humanitarian miracle. But what if the order had been reversed? Is it really possible to develop the countermeasures against CBW without gaining knowledge that could be used aggressively? John von Neumann, despite his misjudgements of other issues, argued persuasively that benign and threatening research could not be neatly separated, observing that scientific and technological branches are linked so closely that only the end of technological progress could hold back potentially dangerous knowledge.

Dilemmas of Success

Even entirely innocuous applications could lead to surprising social changes. Will nano fabrication be controlled, for example, by a small number of patents? If so, who will hold them and how will they be licensed? Will high fixed costs tend to concentrate production in a few global firms, as in the microchip industry, or will there be opportunity for independents alongside the giants? These questions may seem remote now, but are likely to have consequences as dramatic as the effects of personal computing and the Internet.

The benefits of successful nanotechnology could raise living standards globally while creating local crises, for example by slashing the world prices of minerals, gemstones, and other resources that are the mainstays of national and regional economies. Many resource-rich countries are already unstable. And even the United States could be adversely affected; the worldwide spread of nanotechnology processes and skills may reduce the advantages of our agricultural and mineral wealth.

In the West, improvements may have paradoxical consequences. If nanotechnology-based medical tests follow the pattern of previous advances, they may increase anxiety and medical costs by producing large numbers of false positives and requiring further testing. (Risk information does not always improve decisions: a recent Swedish study found that parents, informed that their children had inherited genes associated with early-onset emphysema, were so upset that they began to smoke more rather than quit as the investigators had expected [Brave 2000].) While many cancers and other diseases can be treated effectively if detected early, information about incipient disease or a nontreatable genetic predisposition could actually be a psychological threat to health: "toxic knowledge," some have called it.

Every important new therapy should be reason for rejoicing, but we should not suppose that more effective prevention and treatment will lower medical costs. To the contrary, as health and longevity improve, society pays more for medical care because large numbers of people live to an advanced age and require even more treatment. In 1997, a research group in the Netherlands even found that if all smokers immediately quit, prolonging their lives, medical expenses to society would increase over time. We all hope that nanotechnology will improve the quality as well as the length of life, but because medical advances tend to increase rather than reduce the long-term need for the time of skilled professionals, they are unlikely to be cheap. Already parts of the United States face severe a severe shortage of nurses. Alzheimer's disease threatens to be the epidemic of the twenty-first century, according to some commentators, compounding the need for personnel.

Hidden labor costs of technological breakthroughs are not peculiar to medicine. Software for conventional computer processors now requires ever-larger teams of well-paid programmers writing ever-bulkier code. Nanotechnology may accelerate the increase of processor speed and storage capacity, but even with new programming tools, it may require even more programming time. Certainly the explosion of electronic resources in education has tended to increase rather than reduce costs.

In all fields, the success of new high-technology methods also has the unintended effect of eroding older skills that remain needed if only as backups. Thus many young physicians now are said to be unable to use a stethoscope properly. Many craft and industrial skills are also declining. Yet in the event of major environmental or social disruption in which high technology no longer functions, these abilities would be crucial resources, as paper-based information would be. In fact some traditional skills remain crucially important even for the most sophisticated technology; experimentation in physics and chemistry still depends on machinists, glassblowers, and other master artisans and technicians.

Planning for the Unexpected

While many paradoxes accompany radical innovation, opposing it can bring equally strange results. If we freeze technology, we perpetuate and amplify the environmental and social costs of the status quo, including the degradation of air and water quality and the acceleration of climate change. We are on a technological treadmill. We have to find new ways to do things, and nanotechnology can not be excluded. Very possibly, its greatest benefit may be an apparently humble one, like the reduction of rolling friction with a new generation of bearings. There is a precedent: Edwin Mansfield of the University of Pennsylvania determined that innovations in sewing thread had improved the standard of living more than high-technology devices (Weinstein 1993). By permitting faster sewing machine speeds, it had substantially reduced the cost of clothing. And apparently trivial research can have profound results: the chemistry of magnetic tape was an indirect consequence of German research on binders for the gold particles on the Black Russian cigarettes of the 1920s (Fantel 1987).

Nanotechnology promises new beginnings. Researchers should bring not only a strong ethical sense but awe at the complexity with which the natural and human worlds interact. They should encourage participation both of lay people and of other professionals familiar with long-term as well as immediate risks: public health specialists, conservation biologists, and environmental historians. No advice can be infallible, but nanotechnology researchers have a rare opportunity to avoid or mitigate the kinds of unintended consequences that have accompanied other major innovations. It would be a great mistake to try to avoid all mistakes. But it would be an even greater error to forget the modesty that the history of technology teaches.

References

Brave, Ralph. 2000. "Genome Hell is Revealed in Many Forms." *Baltimore Sun*, July 2, p. 1C.

Fantel, Hans. 1987. "Portable CD Players Advance." *New York Times*, May 17, Section 2, p. 33, col. 1.

National Science and Technology Council (NSTC). 1999. *Nanotechnology Research Directions: IWGN Workshop Report.* (http://itri.loyola.edu/nano/IWGN.Research.Directions/) – also published by Kluwer Academic Publishers, Dordrecht, The Netherlands.

Perrow. C. 1984. *Normal Accidents.* New York: Basic Books.

von Neumann, J. 1956. *The Fabulous Future.* New York: Dutton. (reprinted from *Fortune*).

Weinstein, Michael M. 1993. "A Test You're Apt to Flunk." *New York Times*, March 28, Section 4, p. 14, col. 1.

A CULTURAL ECOLOGY OF NANOTECHNOLOGY

B.A. Nardi, Agilent Laboratories

Radically new technologies imply radically new social issues and opportunities. I propose a cultural ecology of nanotechnology in which we find ways to infuse technological development with deeper, more thoughtful and wide-ranging discussions of the social purposes of technology. I chose the ecology metaphor to signify the integration of science and society, to draw attention to interdependencies characteristic of ecologies. (See Nardi and O'Day 1999 on *information ecologies*.)

As part of a nanotechnology initiative I would like to see a new science of cost-benefit analysis in which issues of ethics and social responsibility as they relate to technology can be rethought in radical new ways. Perhaps the term "cost-benefit analysis" should be replaced as it connotes limited economic considerations to many. We need to channel energy into the invention of a holistic process of technological development within which we can entertain questions of the human purposes and benefits of technology. The thrust of such an effort would not be prediction or a simplistic notion of managed change, but the development of a new way of approaching our relationship to technology.

In a cultural ecology of nanotechnology, we would take seriously the promises of nanotechnology such as cleaner manufacturing, decreased waste, or marvelous medical devices. We would put socially beneficial technologies at the top of the research list. We would find new ways to distribute technologies such as medical devices equitably, we would encourage (somehow) companies to use safer technologies such as, say, nanotech tires that don't fray. Whatever our social desiderata, we would find ways to fuse them to the development and deployment of new technology.

For me, such a process of designing ecologies of technology is desirable because I am not as optimistic as others that technological development always comes out for the best in the long run. Sometimes it does and sometimes it does not. I feel we are currently paying too high a price in pollution, noise, traffic congestion, loss of nature, and lack of safety in our technologies. These poor outcomes are a result of the characteristics of specific technologies coupled with the ways we use the technologies. Traffic gridlock could probably never have happened without the electric self-starter; it is important to remember that specific technologies do matter. On the use side, we needn't have had gridlock had we planned our transportation system differently, insisting on a diverse system of public and private vehicles. If we had gone with the early electric vehicles that were starting to be marketed early in the 20th century (e.g., Figure 6.31), air pollution would be much less of a problem.

319

Figure 6.31. Early electric car.

In a cultural ecology of technology, the relationships between attributes of specific technologies and the ways we actually use the technologies are a key focus of interest. In a cultural ecology of nanotechnology, I'd like to see discussion of how to embed the new technologies into society in radically new, socially beneficial ways. We need to have discussions, for example, about the social contradictions inherent in some of the nanotech promises. How do more long-lasting durable products play with our current economic system based on not-so-durable products that must be replaced often to increase profits for manufacturers? Nanotechnology promises tires that don't fray but we have technologies for safer vehicles now that corporations have sometimes chosen not to use. We know how to manufacture less wastefully, but often we don't do it because it reduces profit. In a cultural ecology of technology such concerns would be a serious topic of discussion and focus of creativity.

The government documents promoting nanotechnology that I have read make no mention of the risks of nanotechnology. Are there none? What if we had had cultural ecologies of technology a hundred years ago, and had thought through the implications of having millions of internal combustion engines on the road? Kettering invented the electric self-starter in 1911. That would have been a great time to undertake a serious envisioning task in which we would have imagined everyone having a car, however fantastic it might have seemed at the time.

Techniques of Envisionment

A key activity of cultural ecologies of technology is envisionment. Techniques for examining multiple possible scenarios exist now and could be expanded and developed. Could we have envisioned millenial Los Angeles in 1911? Probably not. We didn't know how. We still don't. But we should learn how.

Actually, we are already envisioning, but doing a poor job of it. We do not hesitate to conjure wondrous benefits of technology, or, less often, to forecast dystopic visions of technological annihilation. Neither is usually realistic. We need to create new processes to envision both benefits and risks of technology *and the relations between them*. This endeavor in itself is an area for technical creativity. That we have been bad at predicting the future in the past is no reason to avoid this critical task now. If we can talk about creating self-replicating machines out of atoms, we can talk about new techniques for envisioning the consequences of technology. There is no reason we cannot apply our sociological and scientific imaginations to assessing the benefits and risks of technology.

I have noticed that proponents of new technologies often follow a two-part logic in advocating for the development of their technology, however risky it might seem to others. The first part of the argument is the confident prediction of great new benefits. For nanotechnology, we have faster computer chips, very high resolution printers, compact high-volume data storage devices, and new medical technologies such as tiny probes, sensors, drug delivery devices, and ways to regenerate bone and tissue. The second part of the argument, should we pose questions about the potential risks of the technology, is that we don't know how to predict where technology is going. When we stop to ponder potential, even likely, risks of technologies, derisive stories of our poor record of prediction in the past are trotted out.

This duality — prediction of fantastic benefits coupled with an assertion that we cannot predict outcomes — is an unhealthy, illogical combination. There's something wrong when the prediction can only be on one side, when we are promised benefits but not allowed to assess risks. Lured by the promises, in which "objective scientific facts" are often invoked as part of the rhetoric of prediction, we go forward, leaving ourselves powerless to envision and prevent negative consequences.

The Russian psychologist Vladimir Zinchenko posits something he calls "the ideal form" as a crucial aspect of human social and psychological development (Zinchenko 1996). The ideal form is where we want to be. Techniques of envisionment are not simply simulations of predicted outcomes, because they contain an element of social purpose. A cultural ecology of technology envisions ideal forms grounded in realistic assumptions, and suggests desirable paths where choices can be made.

Design for Co-evolution

A second thing to work out in a cultural ecology is design for co-evolution. To do this, we would give ourselves ample time for discussing and designing how a social and a technical process could co-evolve together gracefully and proactively. Possible venues for such discussions are workshops and programs sponsored by agencies such as the National Science Foundation.

The notion of design for co-evolution resonates with the idea of co-evolution in Brown and Duguid's response to Bill Joy's *Wired* article on the dangers of nanotechnology. Brown and Duguid point out the strong social influences on technological development. However, I would like to propose that we design and implement a socio-scientific cycle quite different than Brown and Duguid's, which is largely reactive. Our current model of technological development is full speed ahead — and then slam on the brakes when we get scared. Brown and Duguid provide examples of the application of posthoc corrective measures with technologies such as nuclear power. As they point out, it took "the environmental movement, anti-nuclear protests, concerned scientists, worried neighbors of Chernobyl and Three Mile Island, and NIMBY corporate shareholder rebellions to slow the nuclear juggernaut to a manageable crawl."

Good grief, do we have call out the scientists, investors, tree huggers, and little old ladies in tennis shoes every time? With nanotechnology, genetically modified foods, cloning, and other technologies with global implications looming, we need a better process. In a cultural ecology of technology, we would be proactive about technological development, not reactive. We would shape technology for our own collectively defined purposes and not confine ourselves to mobilizing to slow dangerous or undesirable juggernauts. And while nuclear power is a hopeful example of co-evolution, it also required the sacrifice of thousands of lives and continuing ill health for many thousands more.

While we have successfully held some technologies at bay, other technologies are out of control — such as the automobile. We don't have a healthy ecology for the automobile. Breathing exhaust and spending one's short existence crawling along the freeway is hardly a gift from the gods. It's difficult to say what will happen with genetically modified foods. Despite protests, the brake has not been applied. Half the soybeans in the U.S. and a third of the corn are grown from genetically modified seed. We do not know what ecological or economic effects this will have in the medium to long run.

Less threatening but still a kind of daily water torture are technologies such as phone menus that leave people feeling frustrated and diminished. The automated voice response system my bank uses employs poor concatenation technology, stringing together individually recorded numbers that produce a sing-song-y, barely comprehensible voice response to simple requests for account information. These voice systems are especially difficult for people who are hard of hearing or not native speakers of English. Our ecologies often have a *monocultural* character, serving the single need of profit or so-called efficiency. (I sometimes wonder whose efficiency is served as I fight my way through a maze of key presses in a phone menu.)

Zinchenko (1996) suggests that two of our most human attributes are creativity and the ability to resist. Brown and Duguid are betting on resistance. This is a time-tested strategy, and one advocated by some of the best minds of our time such as Michel Foucault and Jacques Ellul. But with the rapid pace of technological change, resistance may no longer be sufficient. It's looking to me like Star Trek was right, "Resistance is futile." By the time we mobilize to resist, a lot of damage may have been done. We need to anticipate and plan. My suggestion is to apply human creativity to the problem of designing our technologies in a process that marries the social and the scientific, that treats technology systematically, ecologically. *I believe we can draw on deep wells of creativity that we have not tapped to do this.*

Brown and Duguid suggest that Bill Joy's concerns about nanotechnology are distorted by technological tunnel vision. But Joy is far from oblivious to the social. He situates the development of nanotechnology very realistically as a product of "global capitalism and its manifold financial incentives and competitive pressures." Capitalism is a social force more powerful than the "government, courts, formal and informal organizations, social movements, professional networks, local communities, and market institutions" enumerated by Brown and Duguid. Indeed many of these social forms are deeply implicated in capitalism, not outside of it. Forces that can "redirect the raw power of technologies," as Brown and Duguid say, come up against the "manifold financial and competitive pressures" of which Joy speaks. Some parts of government and certain social movements do exist as reactive forces trying to slow and restrain rampant capitalism. However, I believe such forces should be primary generative stimuli of planned societal progress, not catch-up rearguard actions. Joy's fear of "undirected growth" is one to take seriously.

Politically experienced people probably find the idea of creating a new social process that intimately links society and science naive and unwieldy. But if we can manufacture devices where a billionth of a meter is a meaningful measure, there is the possibility that we can shape our social processes in just as radical a way.

References

Brown, J.S. and Duguid, P. 2001. This volume.

Ellul, J. 1964. *The Technological Society*. New York: Alfred Knopf. (First published 1954).

Foucault, M. 1977. Power/Knowledge: Selected Interviews and Other Writings 1972-1977. New York, Pantheon Books.

Joy, W. 2000. Why the Future Doesn't Need Us. *Wired*, April.

Nardi, B., O'Day, V. 1999. Information Ecologies: Using Technology with Heart. MIT Press.

Zinchenko, V. 1996. Developing Activity Theory. In B. Nardi, Ed. *Context and Consciousness: Activity Theory and Human-Computer Interaction.* MIT Press.

ENVISIONING AND COMMUNICATING NANOTECHNOLOGY TO THE PUBLIC

Felice Frankel, MIT

The importance of communicating advances in nanotechnology to the public will be essential if we are to expect intelligent feedback on our efforts. The question lies in how are we to provide information of such a highly technical nature to a public not yet accustomed to the language of phenomena on the nano level.

One way is to use pictures. Images made with a deep and simultaneous respect for the technical, informational and aesthetic content will be important for the communication of nanotechnology to the public.

When used with intelligent and accessible text, images of all kinds (e.g., Fig. 6.32) will play a major role in engaging the interest of the general public. Engagement is the first step in the public's accessibility to information allowing for improved and productive feedback.

Figure 6.32. Polymer rods embedded with cadmium selenide nanocrystals, fluorescing at various wavelengths. Researchers: M. Bawendi, K. Jensen, J. Lee, MIT (copyright Felice Frankel; reprinted by permission).

The following figure (Fig. 6.33) is an example of how a photograph was used to convey similar information as a graph. The numbers below each cuvette on the right correspond to the peaks in the graph showing various wavelengths of photoluminescence on the left.

For general audiences, the photograph is usually less intimidating than the graph and may be one way to permit the viewer to enter into the world of research and to ask questions. There are, of course, a number of other visual representations of data, i.e., animation, illustration, etc. All should be considered part of any investigation of research, especially in nanotechnology. SEEING what are the complex phenomena of nanotechnological research can only encourage expanded interest to the public.

> I believe that we who are privileged to see science's splendor, who image it, diagram it, model it, graph it, and compose its data, can turn the world around, dazzling it with what inspires and nourishes our thinking, if we refine the visual vocabulary we use to communicate our investigations and incorporate — beautifully and above all accurately — the visual component that is already there. Our goal must be to share the visual richness of our world, to make it accessible (Frankel 1998).

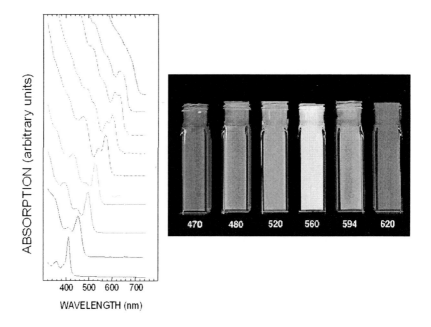

Figure 6.33. Two Representations of Quantum Confinement:
Left - Spectrographic Display (C.B. Murray, MIT), Right - Photograph (F. Frankel, MIT).

References

Frankel, F. 1998. "Essays on Science and Society, Envisioning Science — A Persona Perspective." *Science*, June.
http://www.sciencemag.org/cgi/content/full/280/5370/1698.

Additional References:

Felice Frankel Homepage: http://web.mit.edu/edgerton/felice/

Image and Meaning Initiative Homepage: http://web.mit.edu/i-m/

F. Frankel, G.M. Whitesides. 1997. *On the Surface of Things, Images of the Extraordinary in Science*. Chronicle Books, San Francisco.

BIBLIOGRAPHY

Publications by the National Science and Technology Council

Interagency Working Group on Nano Science, Engineering, and Technology. *Nanotechnology Research Directions*. September 1999. Available at: http://nano.gov/ and published by Kluwer Academic Publishers, 2000. This publication incorporates a vision for how the nanotechnology community — Federal agencies, industries, universities, and professional societies — can coordinate efforts more effectively to develop a wide range of revolutionary commercial applications. It identifies challenges and opportunities in the nanotechnology field and begins to make recommendations on how to develop a balanced infrastructure for nanotechnology R&D, advance critical research areas, and nurture the scientific and technical workforce of the next century.

Interagency Working Group on Nano Science, Engineering, and Technology. *Nanostructure Science and Technology: A Worldwide Study*. August 1999. Available at: http://nano.gov/ and published by Kluwer Academic Publishers, 2000. This report reviews the status of R&D in nanoparticles, nanostructured materials, and nanodevices, including innovative approaches to synthesis and characterization. The report highlights applications in dispersions, high-surface-area materials, electronic and magnetic devices, nanostructured materials, and biological systems. It includes a comparative review of research programs around the world — the United States, Japan, Western Europe, and other countries — to provide a global picture of the field.

Interagency Working Group on Nano Science, Engineering, and Technology. *Nanotechnology: Shaping the World Atom by Atom*. December 1999. Available at: http://nano.gov/. This brochure sets the stage for increasing the public's understanding of what nanotechnology is, how it came to be, and its potential impact on society.

Interagency Working Group on Nano Science, Engineering, and Technology. *National Nanotechnology Initiative – Leading to the Next Industrial Revolution*. February 2000. Available at: http://nano.gov/. This report supplements the President's FY 2001 budget and highlights the nanotechnology funding mechanisms developed for this new initiative and the funding allocations for each participating Federal agency.

Subcommittee on Nanoscale Science, Engineering, and Technology. *National Nanotechnology Initiative: The Initiative and its Implementation Plan.* July 2000. Available at: http://nano.gov/. This report includes the description of the initiative and its implementation plan for eight U.S. Government departments and agencies.

Other Publications

Bimber, B., and Guston, D., eds., *Technology Assessment: The End of OTA*, special issue of *Technology Forecasting and Social Change*, Vol. 54, No. 2 and 3, 1997

Christensen, Clayton M. The Innovator's Dilemma: When New Technologies Cause Great Firms to Fail. New York: Harperbusiness, 2000.

Freitas, L. *Nanomedicine.* Bioscience, 1999.

Joy, Bill. "Why the Future Doesn't Need Us." *Wired*, April 8, 2000. http://www.wired.com/wired/archive/8.04/joy_pr.html/.

Kennedy, Paul. The Rise and Fall of the Great Powers, Economic Change and Military Conflict from 1500 to 2000. New York: Random House,1987.

National Human Genome Research Institute, *Ethical, Legal, and Social Implications of Human Genetics Research of the National Institutes of Health.* Available at: http://www.nhgri.nih.gov/ELSI, October 2000.

National Coordination Office for Information Technology, *Ethical, Legal, and Societal Implications of Information Technology Research and Development.* Available at: http://www.itrd.gov, October 1999.

Smith, A., The Wealth of Nations: An Inquiry into the Nature and Causes, Random House, New York, 1994.

RAND Corporation. The Potential of Nanotechnology for Molecular Manufacturing. 1995.

Winner, Langdon, ed., *Democracy in a Technological Society, in Philosophy and Technology*, Vol. 9, Kluwer Academic Publishers, Dordrecht, The Netherlands, 1992.

APPENDICES

Appendix A. List of Participants and Contributors (as of Sept. 2000)

Government and National Laboratories

William S. Bainbridge
Science Advisor
National Science Foundation
Directorate for Social, Behavioral
& Economic Sciences
4201 Wilson Blvd., Suite 525
Arlington, VA 22230

Joseph Bordogna
Deputy Director
National Science Foundation
4201 Wilson Blvd., Suite 1205 N.
Arlington, VA 22230

Rita Colwell
Director
National Science Foundation
4201 Wilson Blvd., Suite 1205 N.
Arlington, VA 22230

Joanne D. Culbertson
Senior Advisor for Planning and
Evaluation
National Science Foundation
Directorate for Engineering
4201 Wilson Blvd., Suite 505 N
Arlington, VA 22230

Michael Daum
National Economic Council
Executive Office of the President

Michael E. Davey
Analyst in Science and
Technology
Congressional Research Service
(CRS)
CRS-STM-LM413
Library of Congress
Washington, D.C.20540-7490

Mildred S. Dresselhaus
Director, Office of Science
Bldg. FORS, Rm. 7B-058
Department of Energy
1000 Independence Avenue, S.W.
Washington, DC 20585

Murray S. Hirschbein
NASA Headquarters
Building: HQ, Room: 6D70
Washington DC 20546-0001

Tom Kalil
Special Assistant to the President
National Economic Council,
White House
230 OEOB
Washington, D.C. 20502

Richard D. Kelley
U.S. Department of Energy
Materials Sciences Division
SC-13, Rm. F-421
Building: GTN
Washington, DC 20545

Richard D. Klausner
Director
National Cancer Institute
NIH, Building 31, Room 11A-48
31 Center Drive
Bethesda, MD 20892-2590

Annalynn Lacombe
DOT Volpe National
Transportation System Center
55 Broadway, Kendall Square
Cambridge, MA 02142

Elbert L. Marsh
Deputy Assistant Director
National Science Foundation
Directorate for Engineering
4201 Wilson Blvd., Suite 505 N
Arlington, VA 22230

Louis A. Martin-Vega
Assistant Director (Acting)
National Science Foundation
Directorate for Engineering
4201 Wilson Blvd., Rm. 505 N
Arlington, VA 22230

James S. Murday,
Superintendent
Chemistry Division
Naval Research Laboratory
Code 6100
Washington, D.C. 20375-5342

S. Tom Picraux
Director, Physical & Chemical
Sciences Center
Sandia National Labs
P.O. Box 5800
Albuquerque, NM 87185-1427

Gernot S. Pomrenke
Mathematics and Space Sciences
Directorate
Air Force Office of Scientific
Research
Ballston Common Towers II
801 N. Randolph Street, Rm. 732
Arlington, VA 22203

M.C. Roco
Senior Advisor for
Nanotechnology
National Science Foundation
Division of Chemical and
Transport Systems
4201 Wilson Blvd., Suite 525
Arlington, VA 22230

Miron L. Straf
National Science Foundation
Directorate for Social, Behavioral
& Economic Sciences
4201 Wilson Blvd., Suite 525
Arlington, VA 22230

Samuel Venneri
Chief Technologist
National Aeronautics and Space
Administration
Headquarters Building, Rm. 9S13
Washington DC 20546-0001

Gerold Yonas
Principal Scientist
Sandia National Laboratory
P. O. Box 5800
Albuquerque, NM 87185-0839

Academic Contributors

April S. Brown
Georgia Institute of Technology
College of Engineering

John Carroll
MIT Sloan School
50 Memorial Dr.
Cambridge, MA 02142

Michael M. Crow
Executive Vice Provost
Columbia University
305 Low Library MC 4312
535 W. 116th St.
New York, NY 10027

Henry Etzkowitz
Science Policy Institute
State University of New York,
Purchase
735 Anderson Hill Rd.
Purchase, NY 10577

Irwin Feller
Institute for Policy Research and
Evaluation
Department of Economics
N250 Burrowes Building
Pennsylvania State University
University Park, PA 16802

Stephen J. Fonash
Nanofabrication Facility
Penn State University

M. Gregory Forest
Prof. of Mathematics
The University of North Carolina
at Chapel Hill
CB# 3100, South Building
Chapel Hill, NC 27599-3100

Felice Frankel
Project Director, "Envisioning
Science"
MIT
77 Massachusetts Avenue
4-405 , Lab: 13-2050
Cambridge, MA   02139

Hans Glimell
Technology and Science Studies
Göteborg University
Box 700, SE 405 30
Göteborg, Sweden

Evelyn Hu
Center for Quantized Electronic
Structures
Science & Technology Center
University of California
Santa Barbara, CA 93106

Kristina M Johnson
Dean, School of Engineering
Duke University
305 Teer Engineering Library
Box 90271
Durham, NC 27708

Robert Langer
MIT
E25-342
77 Massachusetts Avenue
Cambridge, MA 02139

David A Lavan
MIT
E25-342
77 Massachusetts Avenue
Cambridge, MA 02139

Lester B. Lave
James Higgins Professor of
Economics and Finance,
Professor of Urban and Public
Affairs,
Professor of Engineering and
Public Policy
Carnegie Mellon University
5000 Forbes Avenue
Pittsburgh, PA 15213

J. Christopher Love
Department of Chemistry and
Chemical Biology
Harvard University
12 Oxford St. - Box #362
Cambridge, MA 02138

James Merz
Vice President for Graduate
Studies and Research
University of Notre Dame
208 Hurley Hall
Notre Dame, IN 46556-5641

Daniel Sarewitz
Center for Science, Policy, and
Outcomes
Columbia University
209 Pennsylvania Ave., SE
Washington, DC 2000

Richard Smalley
Chemistry Department
Smalley Research MS100
Rice University
P.O. Box 1892
Houston, TX 77251-1892

Samuel I. Stupp
Northwestern University
Materials and Life Sciences
Building
2225 N. Campus Drive
Evanston, IL 60208

Mark Suchman
The Health Policy Scholars
Program
Yale University
89 Trumbull St.
New Haven, CT 06520

Edward Tenner
Department of Geosciences
Princeton University
Princeton, NJ 08544

Paul B. Thompson
Distinguished Professor
Department of Philosophy
1360 LAEB
Purdue University
West Lafaette, IN 47907-1360

Viola Vogel
Director, Center for
Nanotechnology
Dept. of Bioengineering
University of Washington
231 Wilcox Hall, Box 352125
Seattle, WA 98195

Vivian Weil
Center for the Study of Ethics in
the Professions
Room 102A, Stuart Building
Illinois Institute of Technology
3300 South Federal Street
Chicago, IL 60616-3793

George Whitesides
Harvard University
Department of Chemistry
12 Oxford St., Mallinckrodt 230
Cambridge, MA 02138

Frank A Wolak
Department of Economics
Stanford University
Stanford, CA 94305-6072

Robert Doering
Texas Instruments
P.O. Box 650311
MS 3730
Dallas, TX 75265

Stephen W. Drew
Merck & Co. (retired)
P.O. Box 2000
Rahway, NJ 07065

Private Sector Contributors

John Armstrong
IBM (retired)
PMB 161,
6 University Dr., Suite 206
Amherst, MA 01002-3820.

John Seely Brown
Xerox Palo Alto Research Center
3333 Coyote Hill Road
Palo Alto, CA 94304

James Canton
President, Institute for Global
Futures
2084 Union St.
San Francisco CA

Praveen Chaudhari
IBM Watson Research Center
Yorktown Heights, NY 10598

Marty C. Cornell
Dow Automotive R&D
Dow Chemical
Auburn Hills, MI 48326

Juan M. Garces
Dow Corporate R&D
Building 1776
Midland, MI 48674

Newt Gingrich
American Enterprise Institute (for
Public Policy Research)
atten: Anne Beighey, Project
Director
1150 17th Street, N.W.
Washington, D.C. 20036

Michael J. Heller
Nanogen
10398 Pacific Center Ct.
San Diego, CA 92121

Phil Kuekes
Hewlett Packard Laboratories
3500 Deer Creek Rd.
MS 26U-12
Palo Alto, CA 94304-1126

Martin Stephan Meyer
PO Box 3283
Brighton BN1 4TD
East Sussex, United Kingdom

Bonnie A. Nardi
Agilent Laboratories
3500 Deer Creek Rd., MS 24M-A
Palo Alto, CA 94304

Richard H. Smith, II
4455 Connecticut Avenue
NW Washington, DC 20008

Tom N. Theis
Director, Physical Sciences
IBM Research,
T.J. Watson Research Center

William M. Tolles, Consultant
(Naval Research Laboratory,
Retired)
8801 Edward Gibbs Place
Alexandria, VA 22309

Stan Williams
Principal Laboratory Scientist
Hewlett Packard Laboratories
3500 Deer Creek Rd.
MS 26U-12
Palo Alto, CA 94304-1126

Appendix B. Selected Endorsements of NNI

(see also: http://nano.gov)

Below are NNI endorsements made in 1999 and 2000 by key leaders in universities, industry, trade associations, professional societies and political leaders that underline societal implications of nanoscience and nanotechnology:

> The Semiconductor Industry Association endorses with enthusiasm the establishment of a National Nanotechnology Initiative (NNI). The semiconductor industry has advanced, and continues to advance at a rapid pace, according to Moore's Law, primarily through scaling, continually reducing the physical dimensions of the devices, and structures that make up the chip. This has led to tremendous growth in productivity in nearly all aspects of the economy, since semiconductors are the fundamental building block of information appliances, which enable us to communicate, calculate, and play. However, in time, the dimensions of the devices will approach the atomic scale, the natural province of nanotechnology. We consider basic research in this area crucially important to keep the economic engine moving forward. We will work with the Administration and Congress to assure that this important initiative comes into being.

> **—George Scalise**
> **President, Semiconductor Industry Association**

> As we enter the third millennium, I can't imagine a more important technological initiative to undertake than the National Nanotechnology Initiative. The results forthcoming from such an initiative will transform our lives and transform the very concept of manufacturing in ways that it's hard to fathom at this moment in time. The bringing together of atoms and bits raises many provocative technological and scientific questions. I believe that such a long-term initiative will have short, medium and long term impact and will help stretch the national imagination.

> **—John Seely Brown**
> **Chief Scientist, Xerox Corporation and**
> **Director, Xerox Palo Alto Research Center**

The National Nanotechnology Initiative (NNI) is an extraordinarily important investment in the future strength of America's economy, industrial base, and scientific leadership. Recent scientific and technical advances have made it possible to assemble materials and components atom by atom, or molecule by molecule. We are just beginning to understand how to use nanotechnology to build devices and machines that imitate the elegance and economy of nature. The gathering nanotechnology revolution will eventually make possible a huge leap in computing power, vastly stronger yet much lighter materials, advances in medical technologies, as well as devices and processes with much lower energy and environmental costs. Nanotechnology may well rival the development of the transistor or telecommunications in its ultimate impact. Yet it is the first technological revolution since World War II in which the United States has not had an early commanding lead. We must invest now in the basic scientific and technological research, infrastructure, and young scientists and engineers who will drive this new field and create the industries of the future.

—Charles M. Vest
President, Massachusetts Institute of Technology

The National Nanotechnology Initiative is a big step in a vitally important direction. It will send a clear signal to the youth of this country that the hard core of physical science (particularly physics and chemistry) and the nanofrontiers of engineering have a rich, rewarding future of great social relevance. The coming high tech of building practical things at the ultimate level of finesse, precise right down to the last atom, has the potential to transform our lives. Physics and chemistry are the principal disciplines that will make this all happen. But they are hard disciplines to master, and far too few have perceived the rewards at the end of the road sufficient to justify the effort. The proposed NNI will help immensely to inspire our youth.

—Richard E. Smalley
Gene and Norman Hackerman Professor of Chemistry and Professor of Physics
Rice University Center for Nanoscale Science and Technology

It's hard to think of an industry that isn't likely to be disrupted by nanotechnology.

—David Bishop
Lucent Technologies' Bell Labs

This letter is to acknowledge my full support and endorsement for the National Nanotechnology Initiative. I believe that this initiative is very important for the nation, and will assure our continued leadership position in high technology. The encompassing potential for nanotechnology will help to contribute to improved healthcare for the nation, continue our countries industrial and economic growth, and provide new technical solutions for many environmental problems.

—Michael J. Heller, Ph.D.
Chief Technical Office, Nanogen Inc.

Having represented the pharmaceutical industry in the PCAST review of the applications of Nanotechnology and its role in the future of the U.S. and Global economy, I should like to add my endorsement of the position presented by the Panel on Nanotechnology. Nanotechnology has the potential for several roles in the Health Care arena:

- Reduction of particle sizes of drug substances to enhance oral availability of new drugs and provide mechanisms to enhance the speed of drug development;

- Development of miniaturized drug delivery systems capable of controlling the release of drugs in a more reliable, time-dependent way than is possible with current technology;

- Development of novel diagnostic technologies for evaluation and identification of diseases within the body;

- Development of higher speed, higher capacity IT systems capable of storing and analyzing the massive amounts of data which will become available on patient genetics, and the potential to use this information for targeting the right drug to the right patient.

The potential applications of Nanotechnology are very significant for future health care, and deserves a focused national effort to develop the fundamental physical, chemical and engineering principles which will fuel its development and application.

—Colin R. Gardner, Ph.D.
Vice President, Pharmaceutical Research and Development, Merck

With the future breakthroughs of nanotechnology, we will be able to make things smaller than a few billionths of a meter. The idea of building machines at molecular scale, once fulfilled, will impact every facet of our lives, such as medicine, health care, computer, information, communication, environment, economy, and many more. Nanotechnology will mandate a highly multidisciplinary approach in education and research, cutting across the boundaries of chemistry, biology, physics, materials, and all aspects of engineering. Our campus and industrial partners applaud the foresight of President Clinton and several agencies lead by NSF on this 2001 federal initiative on "nanotechnology." We look forward with excitement to resonating to this challenging initiative.

—**Henry T. Yang**
Chancellor, University of California Santa Barbara

The National Nanotechnology Initiative (NNI) is an important endeavor for this nation to undertake particularly at this juncture of the technology's development. Without question, nanotechnologies will evolve into one of the most significant technological developments of the early 21st Century having major implications in fields ranging from medicine and health, agriculture, electronics, materials science and pharmaceuticals, to name just a few. In the field of semiconductors, the current technology is approaching the point where fundamental changes will be required to enable the industry's continuation down the historical "Moore's Law" path of reducing feature size and cost per bit to achieve continued functional growth — essential to continued productivity gains for the economy. Nanotechnology research represents a promising solution to this challenge, and enables our country to maintain our leadership position in the global high technology race.

Because the foreseeable applications for this technology are perhaps decades away from commercial reality, this basic technology is a classic candidate for federal funding and scientific pursuit. In addition, federal emphasis on nanotechnology through funding grants and scientific research within government, industry and university laboratories would serve to stimulate interest in science and technology among young men and women at a time when there exists a critical need for such resources in the country.

—**Yoshio Nishi**
Senior Vice President, Research & Development, Texas Instruments
Incorporated

There has never been a more crucial time for the U.S. government to support basic research. Besides entering a "Knowledge Age," we are at the threshold of significant discoveries that will return tremendous economic benefits and radically improve every aspect of our lives. Nanotechnology is arguably one of the most promising of these areas, but one that will require long term research across many disciplines to achieve its full promise. Research in nanotechnology will focus the efforts of biologists, chemists, physicists and materials scientists to yield remarkable new materials and devices for medical diagnostics and treatment, computer technology and information management, and technologies for agriculture and energy production.

I wholeheartedly support the National Nanotechnology Initiative (NNI). It's the right approach at the right time, and it provides educational support that will be the lifeblood of our future scientific progress. We must increase funding now for programs such as NNI if we are to maintain technological leadership in the near future.

—Paul Horn
Senior Vice President, IBM Research

The National Nanotechnology Initiative will support atomic, molecular, interface, and nanostructure research applicable to Mississippi State University's strategic research initiatives. These focus on sensor technologies, computational technologies, biotechnologies, and remote sensing, all addressing the agricultural, environmental, and industrial needs in Mississippi.

The sensor miniaturization program in our chemistry department and the simulation, modeling, and visualization capabilities of our Engineering Research Center in Computational Field Simulation provide cross-cutting technologies in support of research and development of sensing systems. These systems have important applications in the areas of environmental pollutants, chemical and biological molecules and proteins, and other nanotechnology structures.

—Malcolm Portera
President, Mississippi State University

Nanoscience and technology research at our Engineering Research Center for Biofilms has been crucial to our understanding weak chemical signals in colonies of bacteria that cause a host of diseases such as middle ear infections, prostatitis, and pneumonia in cystic fibrosis. Greater investment in research at the nanoscale will enable us to work on cures.

—**Tom McCoy**
Acting President, Vice President for Research, Creativity and Technology
Transfer
Montana State University

As Director of Hewlett-Packard Laboratories, I would like to endorse the recommendations of your Committee of Advisors on Science and Technology in their support of the proposed Nanotechnology Initiative. As you may know, HP has been an industry leader in the development of computer technology based upon atomic and molecular structures. The ability to construct machines at the atomic scale will create exciting opportunities for developing new solutions to age-old problems in health and medicine, energy efficiency, agricultural productivity, and in preserving the environment. This development of intelligent, energy efficient and recyclable devices, whose size and weight will be measured in atoms, will likely drive the next wave of economic progress in this country and around the world.

For these reasons, Hewlett-Packard is committed to pursuing opportunities in nanotechnology. We recognize the great opportunities that nanotechnology holds for our country, and therefore we endorse your advisory committee's recommendations that these public policy goals – and their achievement – should be those of the nation as a whole.

—**Dick Lampman**
Director, Hewlett-Packard Laboratories

As the elected representative of the 13,000-member Materials Research Society, I am writing to enthusiastically endorse the National Nanotechnology Initiative. This relatively new and exciting area of science and engineering holds tremendous promise for discoveries and inventions across a wide variety of areas. We see in nanotechnology opportunities for the development of new knowledge, techniques and devices with applications ranging from medicine to computers and telecommunications to aerospace. The ability to control materials near the atomic level to alter properties, tailor their behavior, and to build unseen devices will bring about a revolution that is currently unimaginable. The multidisciplinary nature of nanotechnology is particularly well-recognized by the MRS, in that our members work in cross-disciplinary arenas including biology, biochemistry, solid state physics, materials science, mechanical engineering, and many more. Their work includes much that is already occurring in the fledgling area of nanotechnology, such as biomimetic structures, nano-scale machines and smart materials. It is expected that the National Nanotechnology Initiative will also provide for the education and training in this area of the scientists, engineers, managers, and leaders of tomorrow. As nano-science and engineering is expected to become another fundamental technology, it is vital that we have both the best-trained practitioners and lay citizenry that must participate in making related social decisions.

Please accept our wholehearted support for the National Nanotechnology Initiative. We are looking forward to working with the President and Congress to build a bipartisan effort to make the Initiative a success.

—Harry A. Atwater
President, Materials Research Society

As President of the 42,000-member American Physical Society, I am pleased to endorse the new federal Initiative on Nanotechnology. The Nanotechnology Initiative will take advantage of extraordinary recent developments in the ability of scientists to work with individual atoms, molecules, and electrons. These new capabilities will lead to a deeper understanding of the fundamental physics of novel atomic and molecular systems and, through this understanding, to a greatly enhanced ability to design new materials and devices. The opportunities for understanding the molecular basis for biological processes are especially exciting. I anticipate that the scientific advances arising from this initiative will revolutionize US industries and sustain our nation's remarkable economic development. Physicists will play key roles in all aspects of the Nanotechnology Initiative. In order to inform our members about this new Initiative, the American Physical Society is planning a special plenary session on nanotechnology at our meeting this March.

—James S. Langer
President, American Physical Society

Nanotechnology is clearly a challenging new frontier for industry and industrial R&D, but one that offers unlimited potential for new products, new processes, and new services that will benefit society in ways we can not yet imagine.

—Charles F. Larson
President, Industrial Research Institute

I support the initiative because interest in nanomaterials has been rapidly growing for the past several years. More and more customers are coming to us looking to use our nanopowders to either dramatically improve existing products or create new products using these materials.

A thrust from the federal government is required not only to encourage basic research in nanomaterials to get a good understanding of the basic science issues involved, but also to bridge the gap between science on the one hand, and implementation in the real world on the other.

—Ganesh Skandan
Vice President for R & D, Nanopowder Enterprises Inc. (Small Business)

As the President of the 10,000 member American Ceramic Society, I am writing to you in support of the National Nanotechnology Initiative. We believe that both the infusion of new funding in support of overall research and development activities, as well as this new initiative, will have a critical impact on the nation's economic growth and global leadership role.

The National Nanotechnology Initiative can have an important impact on broad areas of science and technology and can put us in a clear leadership position in this area (something we currently do not hold). Because this is a relatively new area, one can envision the possibility of numerous advances in materials, chemistry, pharmaceuticals, medicine, electronics, information and computer technologies, etc. As with any new research initiatives, one cannot accurately predict specific future breakthroughs. However, we know from past experiences that they are the basis for important new technologies and new civilian markets. One need only review the changes brought about by developments that have evolved from advances in communications and information technology; the space, energy, and nuclear programs; and in the areas of genetics and biotechnology.

The real danger is that of our current situation in which our stagnant support for R&D in the physical sciences is leading to a real decay in our technological and scientific leadership. Last fall,

343

ACerS and many other technical materials societies, whose combined membership represent approximately one million members, wrote to individual congressmen asking them to reconsider the need to increase R&D funding. Prior to that, these technical societies had met with congressional leaders to support R&D funding increases. This continues to be a critical issue for this nation.

We fully support the National Nanotechnology Initiative as an extremely step towards the evolution of new technologies and revolutionary scientific discoveries. We further implore the administration and Congress to work together to increase support for all R&D, which continues to contribute to the growth of our nation's economy and technological and global leadership. We of the American Ceramic Society will strive to provide our support and leadership as we need towards these ends.

—**Paul F. Bechter**
President, The American Ceramic Society

Having first hand research experience in the field of nanotechnology, I would like to endorse the position presented by the panel on nanotechnology. Nanotechnology has excellent potential in revolutionizing Health Care industries. The reduction in size of pharmaceutically active ingredients should increase the stability and bioavailability of the drug. The nanodrug delivery systems will have extraordinary feature such as targeted ultracontrolled release of drugs, vis-a-vis, existing drug delivery systems. The nanobiomaterials have opened new opportunities in designing superior biocompatible coatings for the implants at a molecular level. The National Nanotechnology Initiative (NNI) is a very important investment for future growth of American economy and scientific leadership. We applaud the initiative taken by President Clinton and several other federal agencies on this topic of vital national interest.

—**C. P. Singh, Ph.D.**
President, Nano Interface Technology, Inc. (Small Business)

The Executives Committees of the Division of Materials Physics and the Division of Condensed Matter Physics of the American Physical Society enthusiastically endorse the National Nanotechnology Initiative. We represent approximately 7,500 professional physicists, including many who are leaders in this emerging area of research.

The ability developed over the last decade to manipulate and study materials at the nanometer length scale offers possibilities for advances in science and technology whose potential impact is so vast that we are only just beginning to get a glimpse of it. We are at a time in the development of this technology similar to the early '90's when the Internet was emerging. Nanotechnology offers enormous potential for discovering new fundamental science, for creating new materials with unique and important properties, and for developing new technology. Recent discoveries include an electronic device based on a single molecule, manipulation of biocellular function via synthetic nanocrystal insertion, and nano-scale sensors able to detect environmental conditions with unprecedented accuracy. The impact will be felt in nearly every area of technology, from information storage and processing, to medicine, to remote sensing, to automobiles and telephones. The impact on fundamental science is equally broadbased, from new tools to measure X-rays from distant galaxies to measuring the properties of individual electrons in semiconductors. In addition, we believe that the ability to visualize and manipulate atoms and molecules will capture the public's imagination and inspire a new public commitment to teaching and learning science.

Development of nanoscale science and technology is dependent on progress in an extraordinary wide range of fields, including physics, chemistry, materials science, biology, and engineering. It underlies a new unity in science where progress often depends on a multidisciplinary approach, and where a technological or scientific advance in one field can create extraordinary opportunities in another.

A strong investment by the nation in nanotechnology will lay the intellectual and technical foundation for sustained advances in cutting edge science, innovative technology, and economic competitiveness over the next quarter century. Nanotechnology is the next great frontier, with challenges and opportunities that will extend our reach and enrich our lives. As physicists, we stand ready to work together with other scientists and engineers to develop the promise of nanotechnology. We welcome the scientific challenges and the technological opportunities. We believe that the National Nanotechnology Initiative will bring unprecedented rewards to our society.

—Frances Hellman
Chair, APS Division of Materials Physics
—Richard A. Webb
Chair, APS Division of Condensed Matter Physics

Letter from Dean D. Allan Bromley sent to President Clinton:

Dear President Clinton:

There are few, if indeed any, areas of science or technology that will not be profoundly changed by the introduction of nanotechnology. For this reason, the National Nanotechnology Initiative is of fundamental importance to our economic competitiveness, to our national security, and to the quality of our lives. As yet, we have only glimpsed the dramatic impact that nanodevices can have in extending or repairing deficits in the human senses, in increasing the sensitivity of our measurements, and in expanding the scope and power of both communications and computations. The Initiative is particularly important in that it will build and strengthen the necessary science and technology infrastructure across the U.S. at research centers and institutions, to keep us at the forefront of this vital new technology. We were slow to appreciate its potential and slow to invest in its development. By coordinating fundamental research investments by the more than 15 federal agencies interested in nanotechnology, the Initiative will ensure maximum possible returns in new knowledge and in young minds trained to use that knowledge in innovative and creative fashion for each tax dollar spent.

Nanotechnology is the sixth truly revolutionary technology introduced in the modern world following the Industrial Revolution of the mid-1700s, the Nuclear Energy Revolution of the 1940?s, The Green Revolution of the 1960?s, The Information Revolution of the 1980?s, and the Bio Technology Revolution of the 1990?s.

—D. Allan Bromley
Former Assistant to The President of the United States
for Science and Technology (1989-1993)

Nanotechnology, the science of developing tools and machines as small as one molecule, will have as big an impact on our lives as transistors and chips did in the past 40 years. Imagine highly specialized machines you ingest, systems for security smaller than a piece of dust and collectively intelligent household appliances and cars. The implications for defense, public safety and health are astounding.

—Newt Gingrich
Former Speaker of the U.S. House of Representatives
(*Washington Post*, October 18, 1999, "We Must Fund the Scientific
Revolution")

As the President of The Minerals, Metals, and Materials Society (TMS) I am writing to enthusiastically endorse the National Nanotechnology Initiative. We represent 10,000 professional materials scientists and engineers, some of whom will certainly be leaders in such a national effort. In fact TMS 10 years ago recognized this as one of the potentially most important unexplored frontiers in materials science. As a consequence TMS began holding a series of symposia in the area for university, industry, and government research laboratory scientists to become educated and excited by the possibilities presented by materials having some characteristic nanometer dimensionality. Between then and now exciting "glimpses" of novel properties, unusual property combinations, and new phenomena have been uncovered in such materials. However, much more remains to be answered. For instance, in most cases it is still not known whether the novel properties are a consequence of new physics at the nanometer scale or just the logical extension to small dimensions of large scale phenomena. A focused national effort is just what is required in order for this area to be explored at a faster rate: US industry, national security, and public health can then capitalize on the discoveries of the last 15-20 years, e.g. in taking advantage of findings like the "giant magnetoresistance effect" in nanolayered thin films, which is revolutionizing the magnetic storage industry. Since existing measurement tools are working at their limits to examine such materials, new devices and equipment need to be developed and a large cadre of students need to be trained in their use. Cross-disciplinary research needs to be encouraged and fostered; and seed money for testing innovative ideas needs to be available. A National Nanotechnology Initiative should accomplish this. . . . TMS, a materials society, is wholeheartedly behind this National Initiative because we see this as THE direction of the future in materials research.

—Y. Austin Chang
President, The Minerals, Metals and Materials Society

This new technology is very exciting and might lead to discoveries that will change the way almost everything, from building materials to vaccines to computers, are designed and made. ... Research in nanotechnology is extremely important to future rates of innovation in the country. Innovation is the key to our comparative advantage in the global economy, yet federal investment in the physical sciences that help drive innovation - math, chemistry, geology, physics, and chemical, mechanical, and electrical engineering - are all declining, as are the number of college and advanced degrees in these areas.... It is vitally important that we increase our investment in the physical sciences, including nanotechnology, if we are to see increases in productivity and incomes in the years ahead.

—Senator Evan Bayh (D-IN)
U.S. Senate Science and Technology Caucus Roundtable Discussion, April 5, 2000

Virginia's Center for Innovative Technology (CIT) enthusiastically endorses the National Nanotechnology Initiative (NNI). It is a critical investment in the future strength of America's economy, industrial base, and scientific leadership.

Currently, researchers at universities, companies, and federal labs in Virginia are active in such applications of nanotechnology research as aerospace, biotechnology, communications, electronics, information technology, and advanced manufacturing. The NNI will help Virginia continue to contribute to critical breakthroughs.

Nanotechnology requires a multidisciplinary and collaborative approach in research and industry. Our academic and industrial partners join CIT in expressing the importance of this initiative to the future of the Commonwealth of Virginia and America.

—Anne Armstrong
President, Virginia's Center for Innovative Technology

Nanotechnology sounds like something that's almost science fiction. It's a word that's easy to think is not a real tangible policy. Well, what nanotechnology means to the average citizen is: will it be able to identify a cancer when it's one cell large?

It's the kind of breakthrough that will open doors to science and health research that are closed if we don't invest in nanotechnology. With a surplus and a time of economic well being in the country we have the ability, and I would say we have the obligation, to invest in that kind of forward-looking research. That's why the president put those proposals forward, it's why we're fighting very hard as the Congress considers our budget request for those priorities. We're going to stick to our guns, and we're going to keep insisting on better funding in these areas, because we believe it's very important.

—Jack Lew
Director, Office of Management and Budget
July 5, 2000, National Press Club

Nanotechnology is another very important NSF program. Nanotechnology refers to the ability to manipulate individual atoms and molecules, making it possible to build machines on the scale of human cells or create materials and structures from the bottom up, building in desired properties. Nanotechnology is at an exploratory state. The Nanotechnology Initiative at NSF will fund over 600 projects and 2500 faculty and students, fund 10 large engineering research and materials research centers and 5 university-based research hubs. These efforts will, among other things, help create the knowledge required to address the fast approaching physical limits to semiconductor performance.

—Senator Trent Lott, Senate Majority Leader
Letter to Senate High-Technology Task Force, August 2000

We started the last century with the industrial revolution and ended it with the information revolution. Now, at the beginning of the 21st century, we are on the verge of a new revolution — THE NANOTECHNOLOGY REVOLUTION.

What is nanotechnology? Why do I believe it is the science and technology that will drive the future? Nanotechnology is the science of creating new materials and devices on the atomic and sub-atomic level through the manipulation of individual atoms and molecules. In Nanotechnology, we are poised to take the next major leap into the future where the possibilities are endless.

. . . . . .

Now, the time is right to establish Nanotechnology as an urgent national priority. Last year, President Clinton released a blueprint and a budget for a National Nanotechnology Initiative. This blueprint, created by an inter-agency working group, is one of the least noted and most important documents of the Clinton Administration.

—Senator Barbara A. Mikulski
Wernher von Braun Lecture at NASA Headquarters, June 12, 2000

Appendix C. Index of Authors

Appendix D. Index of Topics